Lecture Notes in Electrical Engineering

Volume 726

The book series *Lecture Notes in Electrical Engineering* (LNEE) publishes the latest developments in Electrical Engineering - quickly, informally and in high quality. While original research reported in proceedings and monographs has traditionally formed the core of LNEE, we also encourage authors to submit books devoted to supporting student education and professional training in the various fields and applications areas of electrical engineering. The series cover classical and emerging topics concerning:

- Communication Engineering, Information Theory and Networks
- Electronics Engineering and Microelectronics
- Signal, Image and Speech Processing
- Wireless and Mobile Communication
- Circuits and Systems
- Energy Systems, Power Electronics and Electrical Machines
- Electro-optical Engineering
- Instrumentation Engineering
- Avionics Engineering
- Control Systems
- Internet-of-Things and Cybersecurity
- Biomedical Devices, MEMS and NEMS

For general information about this book series, comments or suggestions, please contact leontina.dicecco@springer.com.

To submit a proposal or request further information, please contact the Publishing Editor in your country:

China

Jasmine Dou, Editor (jasmine.dou@springer.com)

India, Japan, Rest of Asia

Swati Meherishi, Editorial Director (Swati.Meherishi@springer.com)

Southeast Asia, Australia, New Zealand

Ramesh Nath Premnath, Editor (ramesh.premnath@springernature.com)

USA, Canada:

Michael Luby, Senior Editor (michael.luby@springer.com)

All other Countries:

Leontina Di Cecco, Senior Editor (leontina.dicecco@springer.com)

**** This series is indexed by EI Compendex and Scopus databases. ****

More information about this series at http://www.springer.com/series/7818

Yingquan Peng · Xinyong Dong
Editors

Proceedings of 2019 International Conference on Optoelectronics and Measurement

Springer

Editors
Yingquan Peng
China Jiliang University
Hangzhou, Zhejiang, China

Xinyong Dong
China Jiliang University
Hangzhou, Zhejiang, China

ISSN 1876-1100 ISSN 1876-1119 (electronic)
Lecture Notes in Electrical Engineering
ISBN 978-981-33-4109-8 ISBN 978-981-33-4110-4 (eBook)
https://doi.org/10.1007/978-981-33-4110-4

This Springer imprint is published by the registered company Springer Nature Singapore Pte Ltd.
The registered company address is: 152 Beach Road, #21-01/04 Gateway East, Singapore 189721, Singapore

Contents

Research Progress of Single Photon Avalanche Diode with Low Dark Count Rate

Tong Chu, Guilan Feng, Tianqi Zhao, and Chunlan Lin

Abstract Single photon avalanche diode (SPAD) in Geiger mode is a kind of detector which can detect extremely faint signal, and has been greatly developed in recent years due to the single photon sensitivity, time resolution on the order of picoseconds and high photon-detection efficiency. The device has been widely applied in time-resolved spectrum measurement, quantum imaging and quantum cryptography. The dark count rate (DCR) is one of the key parameters to judge the performance of the device. However, high DCR is a common problem in the world, especially for the device with large sensitive area. In recent years, the research of the device with low DCR is in-depth, and some remarkable achievements have been made, including the establishment of the physical simulation model and the optimization of structure. In this paper, based on the previous research results, we analyze the physical mechanism of the generation of DCR, including Shockley Read Hall (SRH), Trap-Assisted Tunneling (TAT), and Band to Band Tunneling (BTBT), and describe the method of physical model simulation using Technology Computer-Aided Design (TCAD) software. On the other hand, we have studied the structure of the device with low DCR and summarize their advantages, respectively. It is hoped that through in-depth research on the physical model and structure design theory, we can find a way to reduce the DCR of the device, so as to improve its performance.

Keywords SPAD · Dark count rate · TCAD simulation

1 Introduction

Silicon single photon avalanche diodes (Si-SPADs), fabricated in planar technology compatible with CMOS circuits, are the mainstream single photon detector, due to its high performance of high photon detection efficiency (PDE), low dark count rate

T. Chu · G. Feng · T. Zhao (✉) · C. Lin
College of Optical and Electronic Technology, China Jiliang University, Hangzhou 310018, China
e-mail: 183416992@qq.com

Y. Peng and X. Dong (eds.), *Proceedings of 2019 International Conference on Optoelectronics and Measurement*, Lecture Notes in Electrical Engineering 726,
https://doi.org/10.1007/978-981-33-4110-4_1

(DCR) and fast timing response [1]. The Si-SPADs have been widely applied in fluorescence lifetime imaging [2], 3-D imaging [3], and quantum key distribution [4], radar ranging [5] and so on. Combined with thermoelectric cooling, the DCR of the device can be suppressed to <10 Hz. However, the ideal DCR level even in the room temperature is dramatically needed for the application in the quantum integrated optics, such as the quantum computing and SPAD arrays camera. This challenge drives the further research and the improvement of manufacturing process of Si-SPADs in recent years. Thanks to the development of semiconductor process simulation and device simulation tools, the research of semiconductor device is simpler, faster and accuracy. Silvaco Technology Computer-Aided Design (TCAD) provides Athena and Atlas to users which can accurately simulate the process and device characteristics based on physical model, and output some internal parameters such as junction depth, electric filed distribution, avalanche breakdown probability and so on.

We will discuss the situation and prospect of process based on the previous research results in this paper. The generation mechanism and the physical model of DCR will be analyzed in Sect. 2. We will introduce the TCAD simulation software for the Si-SPADs in Sect. 3. In Sect. 4, several device structures with low DCR will be discussed, and their advantages are summarized, respectively. Conclusions will be drawn in Sect. 5.

2 Physical Model of DCR

SPAD operated at excess bias voltage, i.e., the called Geiger mode, where reverse bias voltage is higher than the breakdown voltage. DCR is the frequency of avalanche events that triggered by the electron-hole pairs caused by thermal generation or tunneling effects in depletion layer, when there is no incident photons. There are many factors contributing to the DCR, such as the defects and impurities in crystals and thermal generation. The research on physical model of avalanche diode backtracks to 1960s, including electron (hole) avalanche probability model, Shockley Read Hall (SRH) generation (recombination) model, Trap-Assisted Tunneling (TAT) model, Band to Band Tunneling (BTBT) model and so on, and a set of mature theories has been presented.

There are five DCR generation mechanisms in SPAD which was proposed by Xu et al. [6]. Three among them are dominant, which are trap assisted thermal generation, trap assisted tunneling generation and band to band tunneling generation, respectively. In low electric field, SRH is the dominant source of DCR without tunneling. The SRH model is calculated as following [7]:

$$G_{SRH} = \frac{pn - n_i^2}{\tau_n[p + n_i \exp(\frac{-(E_t - E_i)}{kT})] + \tau_p[n + n_i \exp(\frac{(E_t - E_i)}{kT})]} \tag{1}$$

where, n_i is the intrinsic carrier concentration. n and p are the non-equilibrium electron and hole concentrations, respectively. τ_n and τ_p are the lifetimes of electrons and holes, respectively. E_i and E_t are defined as intrinsic Fermi energy level and recombination center energy level, respectively. k is the Boltzmann constant and T is the absolute temperature.

In strong electric field, electrons can tunnel through the band gap via trap states. The TAT model was analyzed by Hurkx et al. [8, 9], and is described by modifying the SRH mode as following.

$$G_{SRH,TAT} = \frac{pn - n_i^2}{\frac{\tau_n}{1+\Gamma_n}[p + n_i \exp(\frac{-(E_t - E_i)}{kT})] + \frac{\tau_p}{1+\Gamma_p}[n + n_i \exp(\frac{(E_t - E_i)}{kT})]} \quad (2)$$

where, Γ_n and Γ_p are the electron and hole enhancement term for Dirac wells, respectively. We suppose $\Gamma_n = \Gamma_p = \Gamma$, when the electric field strength applied is not more than 9×10^5 V/cm. Γ can be calculated by:

$$\Gamma = 2\sqrt{3}\frac{|E(x)|}{\Gamma_F} \exp((\frac{|E(x)|}{\Gamma_F})^2) \quad (3)$$

$$\Gamma_F = \frac{\sqrt{24 \mathrm{m}_t^* (kT)^3}}{q\hbar} \quad (4)$$

where $E(x)$ is the local electric field strength at depth position x. m_t^* is the effective mass of the tunneling electrons for silicon. q is the electron charge, and $\hbar$ is the reduce Dirac constant.

If the electric field in avalanche region of the device is more than 7 × 105 V/cm, the BTBT contribution is found to be important at room temperature. Rigorous theory and simplified model of the BTBT in silicon was proposed by Schenk and Hurkx et al. [10], and the formula is as following:

$$G_{BTBT}(x) = D \cdot BB.A \cdot |E(x)|^{BB,GMMA} \cdot \exp(\frac{-BB.B}{E(x)}) \quad (5)$$

where D is statistics factor, whose value is set to 1. BB.A, BB.B and BB.GMMA is prefactor, which have two sets of values with direct transition and indirect transition, respectively. The values are set as follows: BB.A = 9.6615 × 1018 cm^{-1} V^{-2} s^{-1}, BB.B = 3×107 V/cm, BB.GMMA = 2, and BB.A = 4×1014 $\mathrm{cm}^{-1}/2$ $\mathrm{V}^{-5/2}$ s^{-1}, BB.B = 1.9 × 107 V/cm, BB.GMMA = 2.5. The latter is chosen in this paper.

In the depletion layer, electron-hole pair will generate an avalanche event according to a certain probability, when the electric field is strong enough. Avalanche probability model was proposed by Oldham and Mcintype et al. [11, 12]:

$$\frac{dP_e}{dx} = (1 - P_e)\alpha_e[P_e + P_h - P_e P_h] \tag{6}$$

$$\frac{dP_h}{dx} = -(1 - P_h)\alpha_h[P_e + P_h - P_e P_h] \tag{7}$$

$$P_{\text{pair}} = P_e + P_h - P_e P_h \tag{8}$$

where, P_e and P_h are the probability of an avalanche event generated by the electron and hole, respectively, and P_{pair} is the joint probability of electron and hole. α_e and α_h are the electron and hole ionization rates, respectively.

The DCR model was proposed by Xu et al. [13] with above mentioned physical models and can be written as:

$$DCR = S \cdot \int_{w1}^{w2} P_{\text{pair}}(x) \cdot (G_{SRH,TAT} + G_{BTBT})dx \tag{9}$$

where S is depletion layer area, *w1* and *w2* are upper and lower boundaries of the depletion layer.

As shown in current results, rigorous theory and formula is proposed so that we can calculate the DCR of SPAD accurately.

3 TCAD Simulation

TCAD has become an vital technology for the development of semiconductor industry. A long cycle and expensive cost is often required if the experimental wafer is used for the research and development of SPAD. While the cost in research and development can be reduced through TCAD simulation. Now we tend to design the device structure and predict the performance with TCAD first. Then, if the performance of device meets the requirement, we will fabricated the device and characterize it.

SILVACO TCAD has become one of the main products in the field of semiconductor process and device simulation, including Athena that provides process simulation, Devedit that provides device editing tools, and Atlas that provides general capabilities for physically-based simulation of semiconductor devices. TCAD simulation takes fully into account the various physical mechanisms including defects and impurities of materials so that it can complete accurate simulation of device characteristics. At present, our research is based on it.

4 Device Structure: Evolution

The structure of SPAD mainly includes a p-n junction in which a avalanche event happened and a guard ring used to avoid premature edge breakdown. First structure of SPAD was introduced by Shockley Laboratory in the early 1960s [14], for the studies of the avalanche multiplication. It's n +/p junction surrounded by a deeply diffused n-type guard ring in p-type substrate, as shown in Fig. 1. This simple structure is operated at low voltage (about 30 V) and fabricated in an ordinary silicon wafer with a planar technology. Nevertheless, it didn't develop due to the limits of monolithic integration technology and silicon foundry services. The epitaxial device with the structure as shown in Fig. 2, was first applied as an efficient photon detectors in [15]. It was fabricated in a p-type epitaxial layer grown on an n-type silicon substrate, which provides a remarkable advantage that a higher quality silicon is used in avalanche region compared with a ordinary silicon substrate.

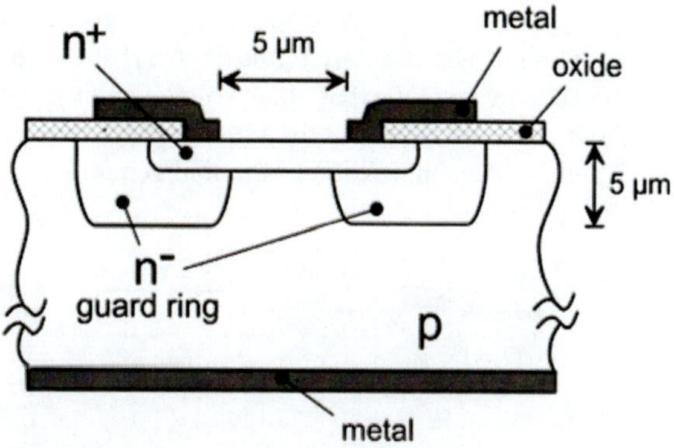

Fig. 1 Cross section of the first structure by Haitz et al. [14]

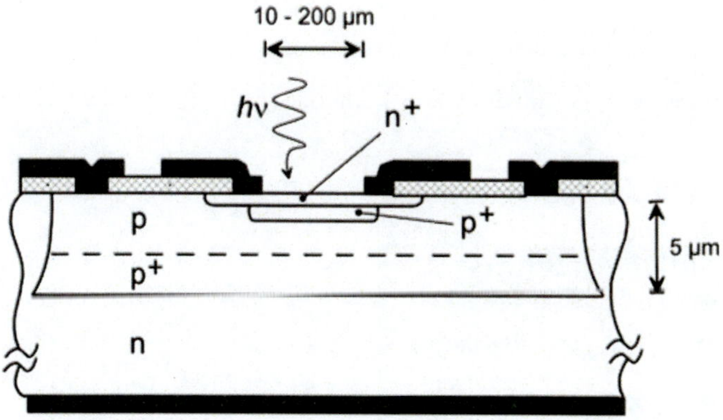

Fig. 2 Cross section of the epitaxial device structure by Lacaita et al. [15]

A new pattern of SPAD doping was proposed by Rochas et al. [16] with the development of high-voltage CMOS technology which provided up to 50 V isolation from the substrate with a low-doped deep n-well. The device consist of a shallow p + and deep n-well junction surrounded by a p-tub guard ring used to avoid premature edge breakdown, as shown in Fig. 3. This structure provided a low DCR and shorter diffusion tail. But a significant shortcoming of HV CMOS SPAD is that depletion layer of junction mostly located in n-well, resulting avalanche event mainly triggered by holes. In silicon, however, holes have a lower probability of avalanche initiation than electrons.

CMOS technology with deep sub-micron (DSM) resolution is mandatory for the fabrication of the SPAD arrays applied in various fields. However, a challenging basic issue must be faced due to the inherent defects of DSM CMOS technologies, such as higher doping concentration in DSM CMOS technology which results BTBT tunneling. A number of structure of SPAD with DSM CMOS technologies have been proposed in recent years [17, 18]. The state-of-the-art SPAD was proposed by Bronzi et al. [19] in 0.35 μm CMOS custom technology where the p-n junction was fabricated through a deep low-doped n-type implant and a shallow high-doped p-type implant. The structure of the device is shown in Fig. 4a. The DCR was below to 0.05 Hz/μm^2 at 6 V excess bias voltage and room temperature. Figure 4b shows the DCR decrease significantly with the work temperature dropping from 50 to − 30 °C, due to the suppression of thermal generation.

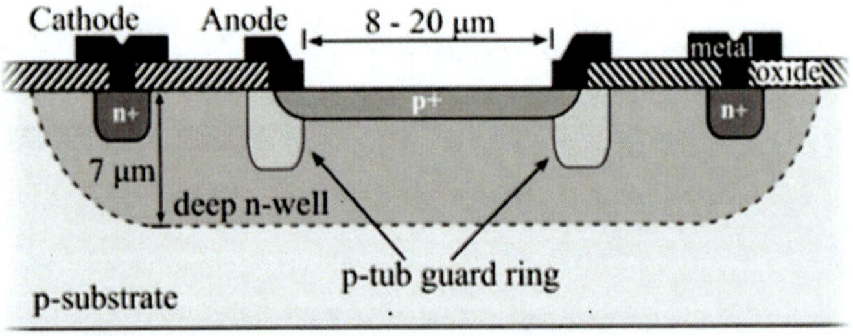

Fig. 3 Cross section of HV CMOS SPAD by Rochas et al. [16]

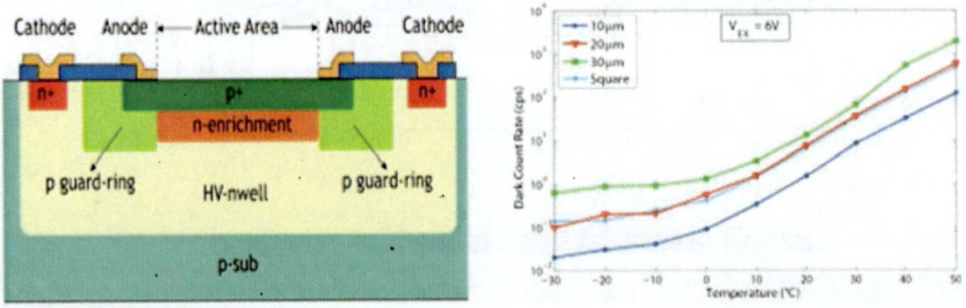

Fig. 4 **a** Cross section of the SPAD by Bronzi et al. **b** DCR of the SPADs, with different area and shapes, as a function of the temperature [19]

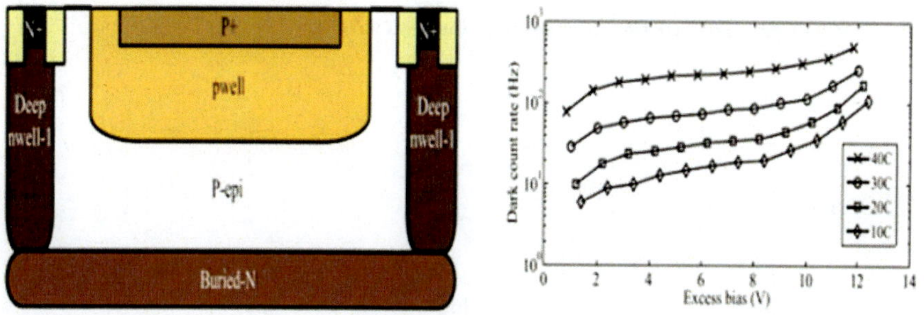

Fig. 5 **a** Corss section of the SPAD by Veerappan et al. **b** DCR of the device performed at various temperature for different excess bias voltages [21]

A virtual guard ring was used all this time until a SiO_2 shallow-trench isolation (STI) guard ring proposed by Finkelstein et al. [20] in DSM CMOS technologies. The new guard ring can be used to withstand the higher electric field and realize a more compactly uniform electric field. A high performance SPAD was proposed by Veerappan et al. with this guard ring [21]. It was fabricated with a p-i-n junction, as shown in Fig. 5a, which resulted in a low electric field strength so that band to band tunneling effect was suppressed. The DCR was below to 1.5 Hz/μm^2 at 11 V excess bias voltage and room temperature. Figure 5b shows minimal growth when the excess bias voltages was limited below 8 V, but significant growth when bias voltages was higher. So it is advisable that the operating bias voltage of the device had better to be less than 8 V.

In order to meet the requirements of the accuracy of the single photon devices, the research of a higher performance SPAD with low DCR must be done. The thermal generation or tunneling effects should be kindly addressed, so that an ideal DCR level can be achieved even at higher excess bias voltage and room temperature.

5 Conclusion

In this paper, physical mechanism of the generation of DCR has been analyzed and TCAD simulation software was briefly introduced. Then, we introduced the evolution of SPAD devices. Two state-of-the-art SPAD devices in different CMOS technologies was analyzed. The SPADs have great development in decades driven by the application of various fields related to single photon detection, such as fluorescence lifetime imaging, 3-D imaging, quantum key distribution and so on. Now, the SPADs with better device performance are highly required, especially for the cutting-edge applications related to the quantum integrated optics, such as the quantum computing and SPAD arrays camera. This article could give some benefits to those who are new in this field.

Acknowledgments This work was supported by the National Nature Science Foundation of China (NSFC) (Grant No. 61904169).

References

1. Ghioni M, Gulinatti A, Rech I, Zappa F, Cova S (2007) Progress in silicon single-photon avalanche diodes. IEEE J Sel Top Quantum Electron 13(4):852–862
2. Lakowicz JR, Szmacinski H, Nowaczyk K, Berndt KW, Johnson M (1992) Fluorescence lifetime imaging. Anal Biochem 202(2):316–330
3. Aull BF, Loomis AH, Youn DJ, Stern A, Zayhowski JJ (2004) Three-dimensional imaging with arrays of geiger-mode avalanche photodiodes. Proc SPIE Int Soc Opt Eng 5353(4):105–116
4. Zbinden H (1998) Quantum cryptography. Appl Phys B 67(6):743–748
5. Schreiber U (1997) Laser radar ranging and atmospheric lidar techniques. Proc SPIE Int Soc Opt Eng 3865(7):1–160
6. Xu Y, Xiang P, Xie X (2016) Comprehensive understanding of dark count mechanisms of single-photon avalanche diodes fabricated in deep sub-micron CMOS technologies. Solid State Electron 129(6):168–174
7. Kindt WJ, Van HW (2002) Modelling and fabrication of geiger mode avalanche photodiodes. IEEE Trans Nucl Sci 45(3):715–719
8. Hurkx GAM, De Graaff HC, Kloosterman WJ, Knuvers MPG (1992) A new analytical diode model including tunneling and avalanche breakdown. IEEE Trans Electron Devices 39(9):2090–2098
9. Hurkx GAM, Klaassen DBM, Knuvers MPG (1992) A new recombination model for device simulation including tunneling. IEEE Trans Electron Devices 39(2):331–338
10. Schenk A (1993) Rigorous theory and simplified model of the band-to-band tunneling in silicon. Solid-State Electron 36(1):19–34
11. Oldham WG, Samuelson RR, Antognetti P (1972) Triggering phenomena in avalanche diodes. IEEE Trans Electron Devices 19(9):1056–1060
12. McIntyre RJ (1973) On the avalanche initiation probability of avalanche diodes above the breakdown voltage. IEEE Trans Electron Devices 20(7):637–641
13. Xu Y, Xiang P, Xie X, Huang Y (2016) A new modeling and simulation method for important statistical performance prediction of single photon avalanche diode detectors. Semicond Sci Technol 31(6):065024
14. Haitz A, Roland H (1965) Mechanisms contributing to the noise pulse rate of avalanche diodes. J Appl Phys 36(10):3123–3131
15. Lacaita MG, Cova S (1989) Double epitaxy improves single-photon avalanche diode performance. Electron Lett 25(13):841–843
16. Rochas A, Gani M, Furrer B, Besse PA, Popovic RS, Ribordy G et al (2003) Single photon detector fabricated in a complementary metal–oxide–semiconductor high-voltage technology. Rev Sci Instrum 74(7):3263–3270
17. Gersbach M, Richardson J, Mazaleyrat E, Hardillier S, Niclass C, Henderson R et al (2009) A low-noise single-photon detector implemented in a 130 nm CMOS imaging process. Solid-State Electron 53(7):803–808
18. Pancheri L, Stoppa D (2011) Low-noise single photon avalanche diodes in 0.15 μm CMOS technology. In: 2011 proceedings of the European solid-state device research conference (ESSDERC), vol 10, number 60. IEEE, pp 179–182
19. Bronzi D, Villa F, Bellisai S, Markovic B, Paschen U (2012) Low-noise and large-area CMOS SPADs with timing response free from slow tails. In: Proceedings of the European Solid-State Device Research Conference (ESSDERC), vol 10, no 63. IEEE, pp 230–233

20. Finkelstei H, Hsu MJ, Esener SC (2006) STI-bounded single-photon avalanche diode in a deep-submicrometer CMOS technology. IEEE Electron Device Lett 27(11):887–889
21. Veerappan C, Charbon E (2016) A low dark count p-i-n diode based SPAD in CMOS technology. IEEE Trans Electron Devices 63(1):65–71

Effects of Temperature on the Performance of Pentcene/SnNc/C_{60} Planar Heterojunction Based Organic Phototransistors

Chengyu Lu, Wenli Lv, and Sunan Xu

Abstract A bottom-gate top-contact OPT based on Pentance/SnNc/C_{60} planar heterojuction was fabricated. The device was characterized at different ambient temperature (T_a) (−40 ~ 110 °C). The dependence of drain current, threshold voltage and field-effect mobi0.lity both in the dark and under illumination on the ambient temperature were revealed. As rising T_a, the drain current increase, and the value of the field-effect mobility (μ) increase first, and then tend to saturate. We suggest that the performance dependence on ambient temperature can be attributed to thermally activated small polaron hopping which facilitate the carrier transport in the enlarged film grains.

Keywords Organic phototransistors · SnNc · Ambient temperature · Field-effect mobility

1 Introduction

In recent years, organic field-effect transistors (OFETs) have received considerable attention because of their applications in the defense construction and civil industry [1, 2]. Among various OFETs, the organic phototransistors (OPTs), which play an important role of photodetectors, have attracted scientists' interests [3].The structure of traditional OFETs were adopted in OPTs, which uses high sensitivity organic materials generally as the active layer. The source, drain, and gate electrodes were optically controlled by light which play a role as an additional terminal [4, 5]. We noticed that nearly all efforts focus on ameliorate high-performance organic electronic devices by searching new materials, developing novel process, and advances in microeconomic theory. In 2006, Takeo Minari et al. have researched a pentacene

C. Lu · W. Lv (✉) · S. Xu
College of Optical and Electronic Technology, China Jiliang University, Hangzhou, People's Republic of China
e-mail: lvwl@cjlu.edu.cn

Y. Peng and X. Dong (eds.), *Proceedings of 2019 International Conference on Optoelectronics and Measurement*, Lecture Notes in Electrical Engineering 726,
https://doi.org/10.1007/978-981-33-4110-4_2

single grain-based OFET about the temperature and electric dependence, they found that the device exhibits a nonmonotonic temperature dependence of the field-effect mobility at 300 to 5.8 K [6]. After that, Joseph A. Letizia et al. [7] reported that the temperature dependence of FET mobility is researched for different channel (p, n-channel), and organic materials based FETs fabricated with varied semiconductor structural in 2010. They found that the carrier mobility of devices are negatively correlated with 1000/T. In this paper, we report the ambient temperature dependence of the electrical performance of OPT based on pentacene/SnNc/C_{60} planar heterojunction. The electrical performance of the device was characterized at different ambient temperature. When the temperature = 80 °C, the drain current and the effect-field mobility reach maximum respectively.

2 Experimental

Pentacene was purchased from Acros Co., SnNc was purchased from sigma-Aldrich Co. and C_{60} from Luminescence Technology Co.. As shown from Fig. 1a, the Si/SiO_2 substrates (gate electrode/gate dielectric) were used for fabricating OPTs with bottom gate and top contact structure. The substrates were cleaned by an Ultrasonic Cleaner in acetone, ethanol and de-ionized water stepwise. And then blow-dried with high pressure N_2 gas and desiccated in a vacuum drying oven (60 °C). For the device fabrication, 50-nm-thick pentacene and SnNc (15 nm)/C_{60} (15 nm) planar heterojunction (PHJ) were deposited stepwise on top of SiO_2. During the whole experiment process, the chamber pressure was kept at 4×10^{-4} Pa and the deposition rate was kept constant at 0.20–0.30 Å/s. Subsequently, Au electrodes (source and drain) were thermally deposited through a shadow mask to control the length and width of the channel (25-μm length (L) and a 2-mm width (W)).

A TU-1901 UV-Vis spectrometer was used for absorption spectra measurements. An organic semiconductor measuring system in a vacuum chamber was used for the measurement of photoelectric properties of devices. The ambient temperature (T_a) was changed in the range from −40 to 110 °C and conducted with a temperature step of 10 °C. In the measurement, first, the T_a reduces from room temperature to −40 °C, and then rises to 110 °C. A 100 mW/cm^2 laser diode with 850 nm wavelength was used for top illumination.

3 Results and Discussion

The UV-Vis absorption spectrum of pentacene, SnNc and C_{60} films on quartz glass are illustrated in Fig. 1b, respectively. One can see that SnNc thin film exhibit strong absorption at the wavelength around 800 ~ 900 nm, while, C_{60} and Pentacene exhibit insignificant ignorable absorption at the same region. As a result, the absorption of Pentance/SnNc/C_{60} PHJ in NIR region originates mainly from SnNc molecule. The

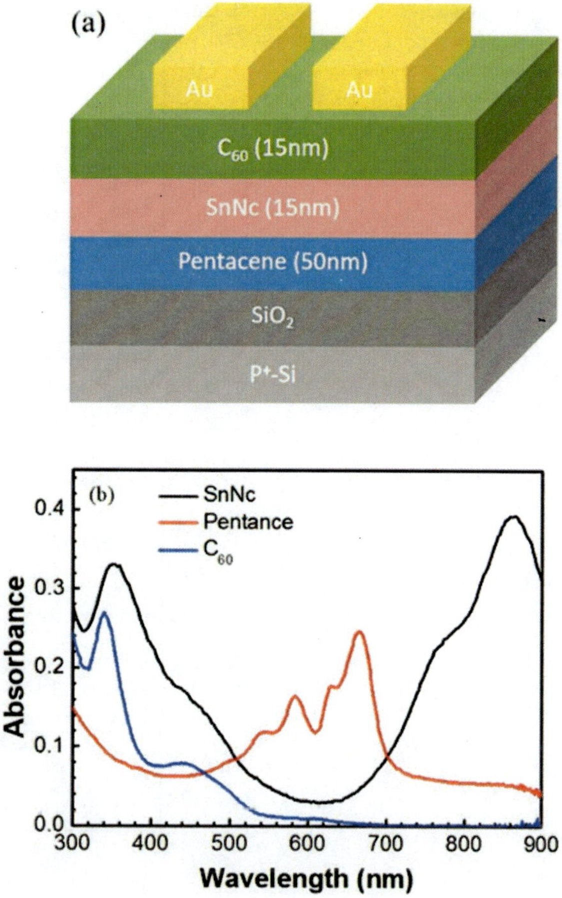

Fig. 1 **a** Schematic structure of pentacene/SnNc/C_{60} OPTs, **b** absorption spectra of SnNc (black line), pentacene (red line) and C_{60} (blue line) thin films

output characteristics of the device in the dark and under illumination at gate voltage (V_g) of −50 V are shown in Figs. 2a and b, typical p-channel operation characteristics are observed. As we can see, the drain current (I_d) of the device decreased with the temperature decreased from 10 to −40 °C in the dark. As the temperature increases from 20 to 110 °C, the current gradually increases. Almost the same phenomenon occurs under illumination with the temperature lowering. However, it can be observed in Fig. 2b that with T_a increasing, the value of saturated zone I_d increases at first, and then decreases after achieving a peak ($T_a = 80$ °C). This is because, as T_a rises for $T_a \leq 80$ °C, the carrier transport was increased by amplified film particles. Subsequently, a decline in I_d with rising T_a for $T_a > 80$ °C is observed. When the temperature over 80 °C, the excessive ambient temperature damages the film's microstructure, result in a poor carrier transport capacity.

For a OPT, the field-effect mobility (μ) and threshold voltage (V_T) at saturation zone are two key performance parameters, and they can be extracted from

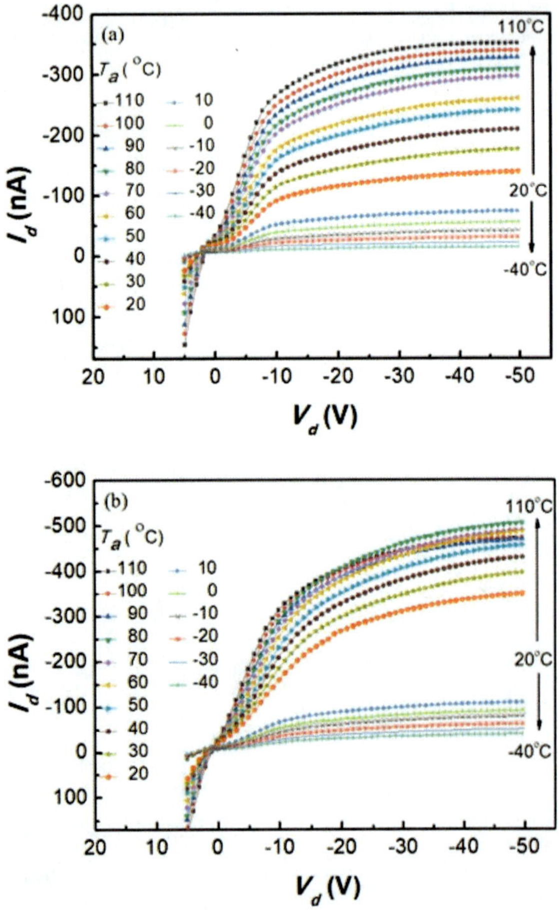

Fig. 2 Output characteristics of the device in the dark (**a**) and under illumination (**b**)

$$I_d = w/2L\, C_i \mu (V_G - V_T)^2 \tag{1}$$

where I_d is the drain current, C_i the gate dielectric capacitance, V_g the gate voltage, and V_d the drain voltage.

The dependence of V_T on the ambient temperature is described in Fig. 3, as T_a increases, the V_T gradually decreases both in dark and under illumination. As T_a decreases, threshold voltage shifts remarkably to the higher potentials. Letizia's theory surmised that the charge between the surface of semiconductor and dielectric is trapped, which shields the gate field [7]. This is similar to our experimental results.

The dependence of field-effect mobility on the ambient temperature is shown in Fig. 4, as we expected, the field-effect mobility curves with a saturation when the ambient temperature increases positively whether in dark or under illumination. According to the Nelson's theory, the increased mobility with T_a can be attributed to

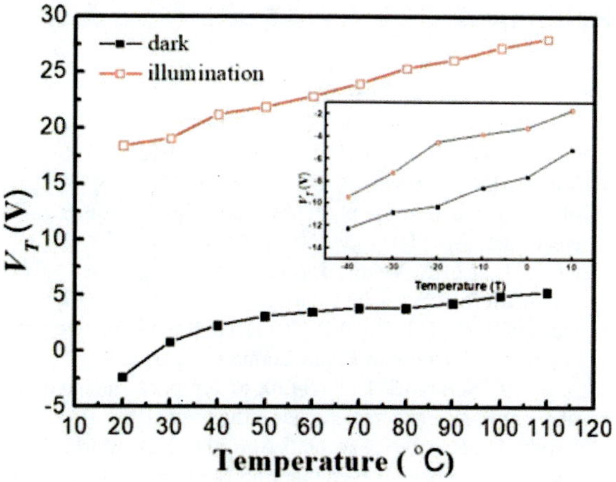

Fig. 3 The dependence of the threshold voltage on the rises ambient temperature from 20 to 110 °C in the dark (filled symbol) and under illumination (empty symbol). The insert displays the trend of threshold voltage with T_a from 10 to −40 °C

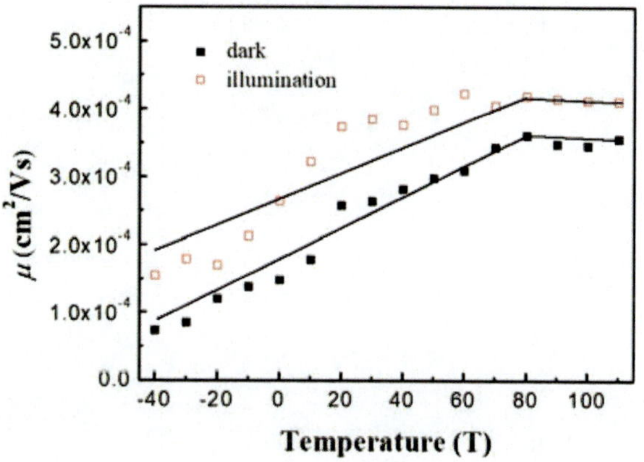

Fig. 4 The dependence of the field-effect mobility (μ) on the ambient temperature in the dark (filled symbol) and under illumination (empty symbol)

the thermally activated small polaron hopping which promotes the motion of carriers through the organic crystal.

4 Conclusion

In this paper, pentacene/SnNc/C_{60} PHJ OPT was fabricated and characterized at different ambient temperature. As T_a rises, the value of V_T and μ of the above-mentioned device increases both in two conditions (dark, illumination). As far as the result concerned, we suggest that the thermally activated small polaron hopping which facilitate the carrier transport in the enlarged film grains.

Acknowledgements This work was supported by the Zhejiang Provincial Natural Science Foundation of China Grant No. LQ19F040003.

References

1. Baeg KJ, Binda M, Natali D, Caironi M, Noh YY (2013) Organic light detectors: photodiodes and phototransistors. Adv Mater 25:4267–4295
2. Guo Y, Yu G, Liu Y (2010) Functional organic field-effect transistors. Adv Mater 22:4427–4447
3. Yao B, Lv W, Chen D, Fan G, Zhou M, Peng Y (2012) Photoresponsivity enhancement of pentacene organic phototransistors by introducing C60 buffer layer under source/drain electrodes. Appl Phys Lett 101:163301
4. Dong HL, Li HX, Wang EJ, Nakashima H, Torimitsu KC, Hu WP (2008) Molecular orientation and field-effect transistors of a rigid rod conjugated polymer thin films. J Phys Chem C 112:19690
5. Dong HL, Bo ZS, Hu WP (2011) High performance phototransistors of a planar conjugated copolymer. Macromol Rapid Commun 32(8):649
6. Minari T, Nemoto T, Isoda S (2006) Temperature and electric-field dependence of the mobility of a single-grain pentacene field-effect transistor. J Appl Phys 99:034506
7. Letizia JA, Rivnay J, Facchetti A, Ratner MA, Marks TJ (2010) Variable temperature mobility analysis of n-channel, p-channel, and ambipolar organic field-effect transistors. Adv Funct Mater 20(1):50–58
8. Nelson SF, Lin Y-Y, Gundlach DJ, Jackson TN, Temperature-independent transport in high-mobility pentacene transistors. Appl Phys Lett 72(15)

A Simple Character Recognition Algorithm on the Image Based on FPGA

Jiayu Song, Peng Wang, Yingquan Peng, and Guipeng Liu

Abstract As we know, the characters recognition uses neural network algorithms typically. However, calculation of neural network algorithms is complex. This paper proposes a new threading algorithm which is suitable for FPGA implementation for the character recognition on an image. The algorithm can identify the characters on an image quickly and accurately. The algorithm selects six lines of image data located the region on the image where the character is located after the image is binarized, and counts the number of "1"s therein, then compares with the parameters gotten by software to obtain the determination result. The accuracy of this algorithm is close to 100%. The algorithm is written on Quartus II software (Version 15.0) of Intel company and implemented on Terasic DE1_SOC. It has the advantages of high reliability and high speed through board level verification and signaltap verification.

Keywords FPGA · Character recognition · Threading algorithm

J. Song · Y. Peng · G. Liu
Institute of Microelectronics, School of Physics Science and Technology, Lanzhou University, 222 South Tianshui Road, Lanzhou 730000, Gansu Province, China
e-mail: yqpeng@cjlu.edu.cn

J. Song · P. Wang · G. Liu
School of Physical Science and Technology, Lanzhou University, 222 South Tianshui Road, Lanzhou 730000, China

P. Wang (✉)
Key Laboratory for Magnetism and Magnetic Materials of the Ministry of Education, Lanzhou University, 222 South Tianshui Road, Lanzhou 730000, China
e-mail: wangpeng@lzu.edu.cn

Y. Peng
Institute of Microelectronics, School of Physical Science and Technology, Lanzhou University, Lanzhou 730000, China

College of Optical and Electronic Technology, China Jiliang University, Hangzhou 310018, People's Republic of China

Y. Peng and X. Dong (eds.), *Proceedings of 2019 International Conference on Optoelectronics and Measurement*, Lecture Notes in Electrical Engineering 726,
https://doi.org/10.1007/978-981-33-4110-4_3

1 Introduction

At present, printed character recognition is closely related to our life and used in various fields [1] widely. Researchers are working to find recognition algorithms that are more accurate and faster. In recent years, a large number of character recognition algorithms have been proposed (e.g. [2–8]). These algorithms mainly include neural network algorithms, pattern matching and threading method. Neural network is a very important identification method, but the programmed is complex and the computation amount is large [9–12], so it is suitable for objects that are difficult to identify. The template matching method is not only computationally intensive, but also has poor anti-interference ability [13]. Thus, in this paper threading method is mainly focuses on the printed character recognition algorithm and the threading method is more suitable for implementation on FPGA [14].

The threading method recognizes characters by extracting the features of the image in the horizontal and vertical directions. Nan Dong uses the threading method to identify the digits displayed on the seven-segment digital tubes on digital instruments; this method is simple and accurate, with almost 100% accuracy [5]. SHEN Xiaoyang obtains the characteristics of each number by selecting 3 vertical lines and 6 horizontal lines in the area of the image where each number is located [15]. However, these threading methods are limited to identifying digits and require precise marking, which places high demands on image positioning and character cutting.

To deal with these problems, this paper proposes an algorithm to extract six region features of the target image to identify the digits and letters on the license plate. The image data collected by the camera is transmitted to the FPGA by pixels. This makes it easy to extract the features of the threading method on the FPGA. At the same time, dividing the target image into six regions to select effective values will improve the accuracy of recognition. All modules in this article were implemented using verilog-2001 and verified on Terasic DE1_SOC.

The paper is organized as follows: Sect. 2 is the implementation principle of the algorithm, while Sect. 3 describes the implementation process of the algorithm on the FPGA, Sect. 4 analyze and demonstrate implementation results. Finally, Sect. 5 gives the conclusions.

2 Algorithm Principle

The character recognition algorithm in this work is called the threading method. The key is to select four regions for the horizontal direction and two regions for the vertical direction of the character image, as shown in Fig. 1, the four regions in the horizontal direction are referred to as Reg_1, Reg_2, Reg_3 and Reg_4, the two regions in the vertical direction are recorded as Reg_5 and Reg_6. The last five characters of the image are the recognition objects (the recognition objects contain digits 0–9 and 24 capital letters, not including the letters I and O). After binarizing

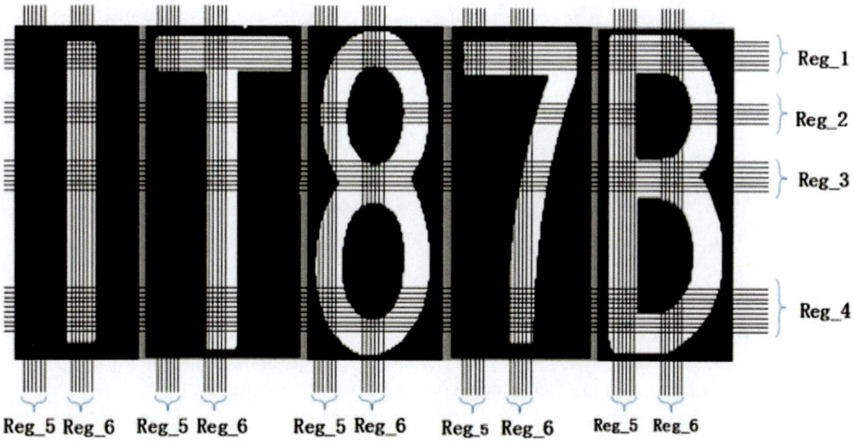

Fig. 1 Locations of six regions on each character

the target image, the image data is only 0 and 1. Count the number of 1 in each row or column in each effective region and use the maximum or minimum value in each region as the effective value of the region. The maximum value of each region is taken as the effective value, and the minimum value in Reg_4 is also taken as the effective value, so each target image will get seven effective values, corresponding to cnt_r1, cnt_r2, cnt_r3, cnt_r4_min, cnt_r4, cnt_r5 and cnt_r6. The next step is to compare these values and their differences with the parameters obtained by software statistics (parameters are expressed in param) to obtain the instruction (inst) for each character.

As shown in Fig. 1, five typical characters those are '1','T', '8', '7', and 'B' are present on the image, the wider the white part of each row or column in the effective region, the larger the effective value will be. The inst of each character is 19 bit (inst [18: 0]), which is obtained through 12 comparisons. The presence of specific characters does not require certain judgments, so the specific bit in these characters inst is marked with x to indicate that this judgment is not affect the result.

The judgment for every bit of inst is present in Fig. 2, when the image width is 806 pixels, the value of param_1 is 15, the value of param_2 is 56, the value of param_3 is 23, the value of param_4 is 5, the value of param_5 is 7, the value of param_6 is 21, the value of param_7 is 154 and the value of param_8 is 52. When the image width changes, these parameters will also be adjusted in proportion.

Finally, the inst of digit '1' are 0001_1010_0100_0xxx_xxx; 0001_1010_1100_0xxx_xxx; 0001_1010_1000_0xxx_xxx; 0001_1011_1000_0xxx_xxx. The inst of the letter 'T' is 1000_000x_xxxx_xxxx_x11. The inst of digit '8' are 1101_1010_111x_xxx1_1xx; 1101_0010_111x_xxx1_1xx; 1101_1010_101x_xxx1_1xx; 1101_0010_101x_xxx1_1xx; 1101_1010_111x_xxx0_1xx; 1101_0010_111x_xxx0_1xx; 1101_1010_101x. The inst of digit '7' is 1000_000x_xxxx_xxxx_x01. The inst of letter 'B' are 1101_1010_111x_xxx1_0xx; 1101_0010_111x_xxx1_0xx;

```verilog
assign   inst[18   ]  = (cnt_r1 < cnt_r2 + param_1)? 1'b0 :  1'b1;
assign   inst[17   ]  = (cnt_r3 < cnt_r2 + param_1)? 1'b0 :  1'b1;
assign   inst[16:14]  = (cnt_r1 > cnt_r3 + param_2)? 3'b000 :
                        (cnt_r1 < cnt_r3 + param_2 && cnt_r1 > cnt_r3 + param_3) ? 3'b001 :
                        (cnt_r1 < cnt_r3 + param_3 && cnt_r3 < cnt_r1 + param_4) ? 3'b011 :
                        (cnt_r3 > cnt_r1 + param_4 && cnt_r3 < cnt_r1 + {param_1,1'b0} )? 3'b010:
                        (cnt_r3 > cnt_r1 + {param_1,1'b0} &&cnt_r3 < cnt_r1 + {param_3,1'b0})?3'b110 : 3'b100;
assign   inst[13:12]  = (cnt_r1 > cnt_r4 + param_2[6:1] ) ?2'b00 :
                        (cnt_r1 < cnt_r4 + param_2[6:1]&&cnt_r4 < cnt_r1 + param_1)?2'b01:
                        (cnt_r4 > cnt_r1 + param_1 && cnt_r4 < cnt_r1 + param_2[6:1] )? 2'b11 : 2'b10;
assign   inst[11:9]   = (cnt_r2 > cnt_r4 + param_2[6:1] )?3'b000:
                        (cnt_r2 < cnt_r4 + param_2[6:1] && cnt_r2 > cnt_r4 + param_5 )?3'b001:
                        (cnt_r2 < cnt_r4 + param_5 && cnt_r4 < cnt_r2 + {param_6,1'b0})?3'b011:
                        (cnt_r4 > cnt_r2 + {param_6,1'b0} && cnt_r4 < cnt_r2 + {param_1,2'b0}) ? 3'b010:3'b110;
assign   inst[8  ]    = (cnt_r2 < {param_1,1'b0}) ? 1'b0 : 1'b1;
assign   inst[7  ]    = (cnt_r4 < {param_1,1'b0})  ? 1'b0 : 1'b1;
assign   inst[6  ]    = (cnt_r3 < {param_4,3'b000})  ? 1'b0 : 1'b1;
assign   inst[5  ]    = (cnt_r4 < param_2)           ? 1'b0 : 1'b1;
assign   inst[4  ]    = (cnt_r4_min < {param_1,1'b0})? 1'b0 : 1'b1;
assign   inst[3:2]    = (cnt_r5 < {param_1,3'b00})?2'b00 :
                        (cnt_r5 > {param_1,3'b00} && cnt_r5 < param_7 - {param_5,1'b0}) ? 2'b01 :
                        (cnt_r5 > param_7 - {param_5,1'b0} && cnt_r5<param_7 + {param_5,1'b0})?2'b11:2'b10;
assign   inst[1:0]    = (cnt_r6 < param_8) ? 2'b00 : (cnt_r6 > param_8 && cnt_r6 < param_7 + param_5) ? 2'b01 : 2'b11;
```

Fig. 2 The code of verilog implements the recognition for each bit of inst

1101_1010_101x_xxx1_0xx; 1101_0010_101x_xxx1_0xx. Although the above-mentioned characters corresponding instructions may be multiple, one instruction corresponds to only one character, thereby ensuring the accuracy of the recognition result.

3 Implementation on FPGA

The whole system is composed of three main modules. As shown in Fig. 3, Mod_1 is an image acquisition module based on OV5640 camera. The function of this module is to collect image data and send it to FPGA; Mod_2 is an algorithm implementation module, the main function is image preprocessing and algorithm implementation. Mod_3 is the image and result display module, which is composed of VGA display and LCD1602 display screen. The function of this module is to verify the accuracy

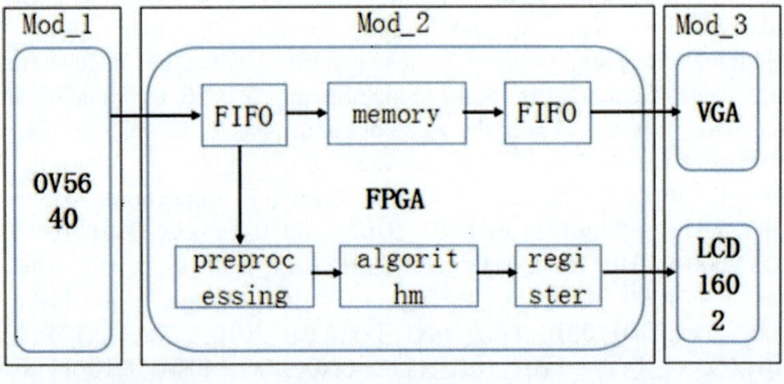

Fig. 3 Hardware system connection diagram: Mod_1 is camera; Mod_2 is connection diagram of each module in FPGA;Mod_3 are the displaying module

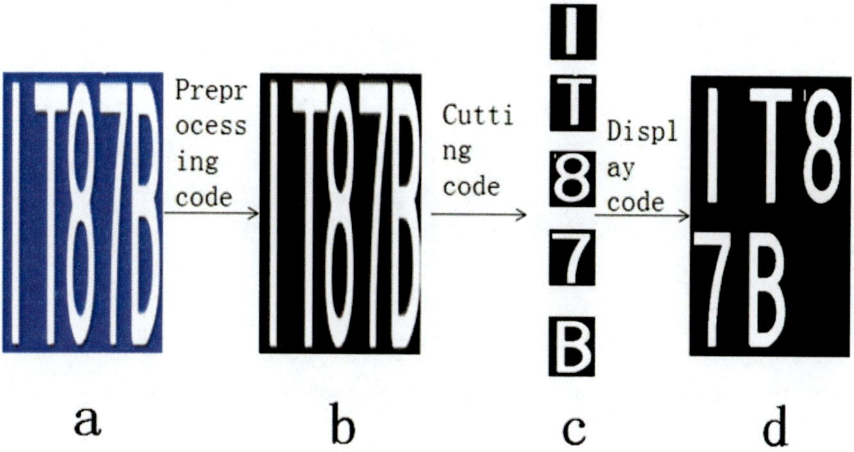

Fig. 4 A typical image displays on VGA: **a** the original image; **b** the binarized image; **c** the digits cut from binarized image; **d** all characters display on VGA

of the image acquisition and recognition results. As shown in Fig. 3, the entire circuit has two paths, the first path is OV5640 to VGA, this way displays the acquisition results; the second path is OV5640 to LCD1602, this way displays the recognition results.

The processing of image is as follows: in the "preprocessing" module, the image is grayed and binarized. Since the license plate characters are white and the rest are blue, the red component data of the image data can be directly used as the gray scale data, which makes it easier to select the binarized threshold. After binarization, the blue background of the license plate becomes black, and the characters become white. Because the upper part of the license plate has longer blue and white borders, it can be used as a criterion for determining the location of the license plate and the size of the license plate. This paper mainly discusses the recognition of 34 characters. The binarized data is sent to the algorithm module through FIFO. Through the VGA display, you can observe whether the original image collected by the camera, the preprocessed binary image, and each character divided are correct. Each character is displayed on VGA is shown in Fig. 4.

4 Discussion and Experimental Results

In the course of the experiment, we obtained the effective value of each character in each effective region through signaltap on Quartus II software, which can effectively verify whether the algorithm proposed mentioned above is correct. As shown in Fig. 5, the effective values and inst of five characters are collected through signaltap, and the effective values of each character are recorded in Table 1, and each value

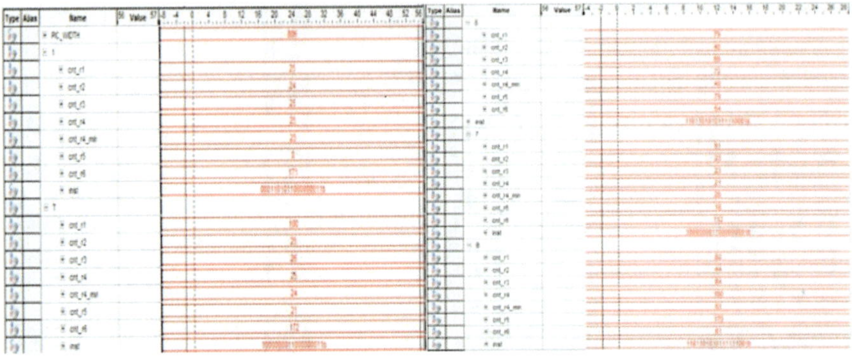

Fig. 5 The valid values and instructions of each character collected by signaltap

Table 1 Valid values of each character

	cnt_r1	cnt_r2	cnt_r3	cnt_r4	cnt_r4_min	cnt_r5	cnt_r6
1	25	24	25	25	23	0	171
T	100	25	26	25	24	21	172
8	79	40	69	72	40	79	54
7	91	23	23	21	20	18	112
B	82	44	84	100	81	170	61

is as expected. The inst collected by signaltap for digit '1' is 0001_1010_1100_0000_011; the inst for letter 'T' is 1000_0000_1100_0000_011; the inst for digit '8' is 1101_1010_1011_1110_001; the inst for digit '7' is 1000_0000_1100_0000_001; the inst for letter 'B' is 1101_1010_1011_1111_001, which are the same as the previous insts and as shown in Fig. 6. The recognition result displayed on the

Fig. 6 Displaying recognition results of the image on LCD1602

LCD1602 is consistent with the characters on the image, indicating that the algorithm is completely accurate on the FPGA and the desired result is achieved.

5 Conclusion

We propose a character recognition algorithm for the image. This algorithm can extract the key features located in the six regions of the image in order to accurately recognize characters, and successfully implemented on the FPGA development board. The algorithm can recognize the results in real time during image acquisition. In the experiment, the captured image size is 1024 * 720, so the time for recognition is 0.0176 s, which is mainly limited to the data collected by the camera and transmitted to the FPGA, and the image data does not need to be cached. (The clock frequency used by the OV5640 is 42 MHz, which is 1024 *720/(4.2* 107) s). Therefore, when the image size remains constant, the recognition speed will increase with the increase of the camera clock frequency, and the recognition accuracy reaches to 98%. The algorithm shows potential applications in number and letter recognition, and we should expect to further improve the accuracy on either number or letter recognition.

References

1. Qin X, Tao Z, Wang X, Dong X (2010) License plate recognition based on improved bp neural network. In: 2010 international conference on computer, mechatronics, control and electronic engineering, vol 23. IEEE, pp 171–174
2. Weng Y, Xia C (2019) A new deep learning-based handwritten character recognition system on mobile computing devices. Mobile Netw Appl 21:1–10
3. Gan J, Wang W, Lu K (2020) Compressing the CNN architecture for in-air handwritten Chinese character recognition. Pattern Recogn Lett 129:190–197
4. Kim HH, Park JK, Oh JH, Kang DJ (2017) Multi-task convolutional neural network system for license plate recognition. Int J Control Autom Syst 15(6):2942–2949
5. Dong N, Zhang Y, Zhao Q (2018) Digital recognition based on sewing method. In: 2018 international conference on advanced mechatronic systems (ICAMechS), vol 33, pp 242–245
6. Xiaoyuan W, Jianping W, Hongfei W (2018) Research on intelligent and digital recognition system and character recognition of electrical instruments. In: 2018 IEEE international conference on mechatronics and automation (ICMA), vol 21, pp 51–56
7. Jian-ning H, Ming-quan W (2010) Research on digital image recognition system based on multiple invariant moments theory and BP neural network. In: 2010 2nd international Asia conference on informatics in control, automation and robotics (CAR 2010), vol 56, pp 399–403
8. Zhang C, Wang Y, Guo J, Zhang H (2010) Digital recognition based on neural network and FPGA implementation. In: 2017 9th international conference on intelligent human-machine systems and cybernetics (IHMSC), vol 27, pp 280–283
9. Chen H, Wang X, Xu B (2016) Study on digital display instrument recognition for substation based on pulse coupled neural network. In: 2016 IEEE international conference on information and automation (ICIA), vol 23, pp 1801–1806

10. Guan T, Liu P, Zeng X, Kim M, Seok M (2019) Recursive binary neural network training model for efficient usage of on-chip memory. IEEE Trans Circuits Syst I Regul Pap 66(7):2593–2605
11. Chang L, Ma X, Wang Z, Zhang Y, Xie Y, Zhao W (2019) Pxnor-bnn: in/with spin-orbit torque mram preset-xnor operation-based binary neural networks. IEEE Trans Very Large Scale Integ (VLSI) Syst 27(11):2668–2679
12. Chiu PF, Choi WH, Ma W, Qin M, Lueker-Boden M (2019) A binarized neural network accelerator with differential crosspoint memristor array for energy-efficient MAC operations. In: 2019 IEEE international symposium on circuits and systems (ISCAS). vol 21, pp 1–5
13. Bin M, Xiangbin M, Xiaofu M, Wufeng L, Linchong H, Dean J (2010) Digital recognition based on image device meters. In: 2010 2nd WRI global congress on intelligent systems, vol 21, pp 326–330
14. Jingbin LU, Li XU (2017) Digital-display instrument recognition system based on adaptive feature extraction. Modern Electron Tech 23:34–41
15. Shen XY, Tang YJ, Jiang BJ, Zhu WL, Sui CH (2013) Numeral characters recognition system of instrument. Instrum Tech Sens 6:23–32

Electromagnetic Line-Parameters Extracted from Microstrip Lines with Step Discontinuities

Zhen Xiao, Dan Zhang, and Weijie Xu

Abstract Microstrip discontinuities are the most fundamental component of MMIC. The study about the accurate modeling of Microstrip discontinuities using different numerical approaches is the most important topics. The electromagnetic waves propagating along microstrip lines with step discontinuities are analyzed using the FDTD method. The results are used in model equations for inhomogeneous transmission lines to extract the equivalent line parameters.

Keywords Generalized transmission line equation · Microstrip line · Distributed parameter

1 Introduction

Recently, the radio communication systems such as cellular phone, PHS, radio-LAN and IC-card are spreading promptly. MMIC has been the center of attention in the capacity of key technology for realizing low-priced and small-sized communication apparatuses. Microstrip discontinuities are the most fundamental component of MMIC. The study about the accurate modeling of Microstrip discontinuities using different numerical approaches is the most important topics. And several papers have been announced in some magazines in the latest years [1–3].

The reason of the electromagnetic noise and the electromagnetic interference according to an imported radio wave in the electronic apparatus could be considered as combining the print-wiring electric signal with the electromagnetic wave [4], for which the systematic description of the equations have not been found now. The coupled problem based on electromagnetic theory could be formulated by taking the current distribution to unknown functions for a simple-structural wiring. But

Z. Xiao · D. Zhang (✉) · W. Xu
College of Information Science and Technology, Nanjing Forestry University, Nanjing 210037, China
e-mail: zhangdan@njfu.edu.cn

Y. Peng and X. Dong (eds.), *Proceedings of 2019 International Conference on Optoelectronics and Measurement*, Lecture Notes in Electrical Engineering 726,
https://doi.org/10.1007/978-981-33-4110-4_4

when the construction became complicated, it cannot be applied easily. While the uniform transmission line equations based on the circuit theory could be considered perspicuity equations for the description that electric signal is transmitted on line using voltage, current and per-unit length inductance (L), capacitance (C), resistance (R) and conductance (G). But when the different constant lines connected, it would be simply treated a connection of the lines which the characteristic impedance is different. The effect and the frequency dependence are not considered that is generated by the high frequency and the non-uniform distributed line constant in the connected segments. The new circuit equations have be constructed that the above-mentioned electromagnetic phenomena on the discontinuities have been taken in the parameters, according to the perspicuity of the distributed constant transmission line equations. However we need to abstract the equation parameters by numerical method, such as the finite difference time domain method (FDTD). In this paper a uniform, step discontinuities parameters have been abstracted as examples [5].

2 Formulation of the Problem

A schematic of a lossless microstrip line considered here is shown in Fig. 1. The microstrip is located on a substrate with a height h and a dielectric constant ε_r and comprises a step discontinuity of a width ω_2 and a length l flanked on each side by a section of uniform lines with a width ω_1. The line is excited by a dominant mode of the uniform transmission line. The electric and magnetic fields propagating along the line is first computed using the FDTD method [6]. To reduce the size of the computational domain, a finite section of the non-uniform line is enclosed by a rectangular parallelepiped space truncated by the PML as shown in Fig. 2, except for the plane $z = 0$ where the condition of perfect electric conductor is implemented. The calculated electric magnetic fields in the time domain is transformed into those in the frequency domain using the discrete Frourier transform technique. Next, the current and voltage along the line is defined by the following equations:

Fig. 1 Voltage and current in strip-line

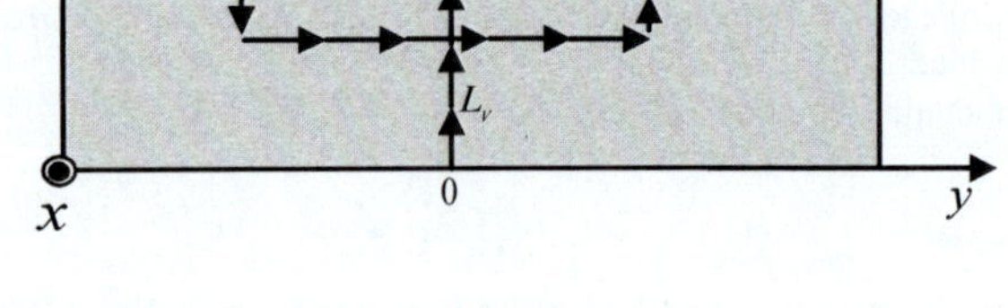

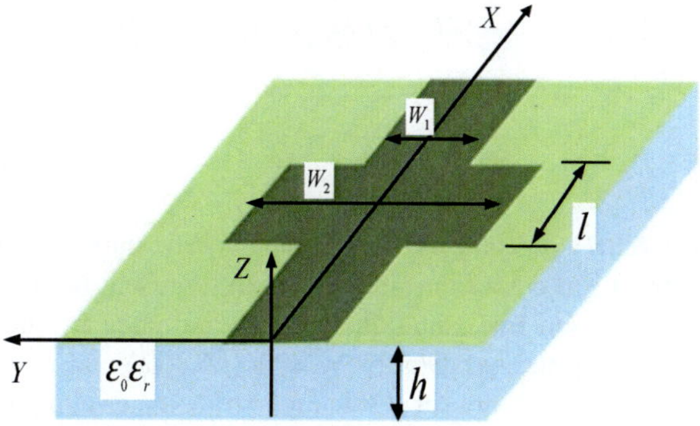

Fig. 2 Step microstrip line structure

$$I(x, \omega) = \oint_C \mathbf{H}_t(x, y, z, \omega) \cdot d\mathbf{s} \tag{1}$$

$$V(x, \omega) = -\int_0^h E_z(x, 0, z, \omega) dz \tag{2}$$

where C_s is a contour enclosing the microstrip on the $y - z$ plane as shown in Fig. 2 and the line voltage is defined in terms of the vertical component of the electric field underneath the microstrip. Since $I(x, \omega)$ is defined a half cell away from where $V(x, \omega)$ is computed, an interpolation scheme is applied to obtain $I(x, \omega)$ at the same nodal points as $V(x, \omega)$.

Finally, the calculated $I_i(\omega) = I(i\Delta x, \omega)$ and $V_i(\omega) = V(i\Delta x, \omega)$ at each nodal point $x = i\Delta x$ are substituted into the transmission line equations to extract the equivalent circuits parameters, where Δx denotes the mesh step-size in the x direction. Although different versions of transmission line equations may be tested, we discuss here two models of the equations. One is the conventional non-uniform transmission line equations given as follows:

$$\frac{\partial}{\partial x} V(x, \omega) = -i\omega L(x, \omega) I(x, \omega) - R(x, \omega) I(x, \omega) \tag{3}$$

$$\frac{\partial}{\partial x} I(x, \omega) = -i\omega C(x, \omega) V(x, \omega) - G(x, \omega) V(x, \omega) \tag{4}$$

where $L(x, \omega)$ and $C(x, \omega)$ are the series inductance and shunt capacitance per unit length along the line and $R(x, \omega)$ and $G(x, \omega)$ are the series resistance and shunt conductance per unit length. Note that since the transmission line is assumed to be lossless, $L(x, \omega)$ and $C(x, \omega)$ take real values, and $R(x, \omega)$ and $G(x, \omega)$ represent the local radiation effects due to the discontinuities. The other is the novel equations which were recently proposed by Zhong, Liu, and Mei [7] as follows:

$$\frac{\partial}{\partial x}V(x,\omega) = -i\omega L(x,\omega)I(x,\omega) + \alpha(x,\omega)V(x,\omega) \tag{5}$$

$$\frac{\partial}{\partial x}I(x,\omega) = -i\omega C(x,\omega)V(x,\omega) + \beta(x,\omega)I(x,\omega) \tag{6}$$

where $\alpha(x, \omega)$ and $\beta(x, \omega)$ are the coefficients for series voltage source and shunt current source per unit length which represent the local radiation effect. The set of transmission line equations are approximated by the finite-difference expressions and the numerical data for $V_i = V(i\Delta x, \omega)$ and $I_i = I(i\Delta x, \omega)$ computed by the FDTD analysis are substituted to extract the circuit parameters of the ith segment of the inhomogeneous line. For the conventional transmission line Eqs. (3) and (4), we have [8]

$$L_i(\omega) = -\,\mathrm{Im}\left[\frac{V_{i+1} - V_{i-1}}{\omega I_i \Delta x}\right] \tag{7}$$

$$R_i(\omega) = -\,\mathrm{Re}\left[\frac{V_{i+1} - V_{i-1}}{I_i \Delta x}\right] \tag{8}$$

$$C_i(\omega) = -\,\mathrm{Im}\left[\frac{I_{i+1} - I_{i-1}}{\omega V_i \Delta x}\right] \tag{9}$$

$$G_i(\omega) = -\,\mathrm{Re}\left[\frac{I_{i+1} - I_{i-1}}{V_i \Delta x}\right] \tag{10}$$

On the other hand, the transmission line Eqs. (5) and (6) lead to

$$L_i(\omega) = -\frac{\mathrm{Im}\left[(V_{i+1} - V_{i-1})V_i^*\right]}{\mathrm{Re}\left[\omega V_i I_i^* \Delta x\right]} \tag{11}$$

$$\alpha_i(\omega) = \frac{\mathrm{Re}\left[(V_{i+1} - V_{i-1})I_i^*\right]}{\mathrm{Re}\left[V_i I_i^* \Delta x\right]} \tag{12}$$

$$C_i(\omega) = -\frac{\mathrm{Im}\left[I^*{}_i(I_{i+1} - I_{i-1})\right]}{\mathrm{Re}\left[\omega V_i I_i^* \Delta x\right]} \tag{13}$$

$$\beta_i(\omega) = \frac{\mathrm{Re}\left[V_i^*(I_{i+1} - I_{i-1})\right]}{\mathrm{Re}\left[V_i I_i^* \Delta x\right]}. \tag{14}$$

where $Re[\;]$ and $Im[\;]$ indicate the real and imaginary parts of variables in the bracket and the asterisk denote the complex conjugate. It should be noted that since the sampled $V_i(\omega)$ and $I_i(\omega)$ are obtained as complex-valued data, the circuit parameters of the lossless transmission lines $L_i(\omega)$ to $\beta_i(\omega)$ can be extracted through a single rum of the FDTD simulation.

3 Numerical Examples

In this paper, we have analyzed the following example. In example, the transmission line is assumed to be lossless. The parameters of the structures used by FDTD are as follows:

Calculation domain:

$$x \times y \times z = 52.98 \times 5.88 \times 2.64\text{mm}^3$$

Space interval: $c_0\Delta t = 0.0173\text{mm}$

$$\Delta x = \Delta y = \Delta z = 1.734 c_0 \Delta t$$

Thickness of PML: 0.24 mm
Reflection error: -120 dB
Pulse excitation:

$$E_{z0}{}^{\text{in}}(x, y, z, t) = \begin{cases} A_0 e^{-(t-T_1)^2/2T_2{}^2} \sin 2\pi f_0 t \delta(x - x_0) \\ 0 \end{cases} \tag{15}$$

$$x_s = 0.57\text{mm} \quad f_0 = 30\text{GHz} \quad T_1 = 72ps \quad T_2 = 24\text{ps} \tag{16}$$

The example is a step microstrip line shown in Figs. 1 and 2. For the structure, relative dielectric constant ε_r is 9.6, height h between metal strip and metal ground plate is 0.6 mm, thickness t of the metal strip is zero, width w of the uniform metal strip is also 0.6 mm. The incident power, reflected power, transmitted power and radiated power of the lines can be obtained using voltage and current or electric and magnetic field given by the FDTD according to the following expressions [9–11]:

Incident power:

$$P_i(\omega) = \frac{1}{2} \text{Re}\{V_i(x, \omega) I_i{}^*(x, \omega)\} \tag{17}$$

Reflected power:

$$P_r(\omega) = \frac{1}{2} \text{Re}\{V_r(x, \omega) I_r{}^*(x, \omega)\} \tag{18}$$

Transmitted power:

$$P_t(\omega) = \frac{1}{2} \text{Re}\{V_t(x, \omega) I_t{}^*(x, \omega)\} \tag{19}$$

Radiated power:

$$P_s(\omega) = \oint_s \frac{1}{2} \mathrm{Re}\{E(\omega)H^*(\omega)\}dS \tag{20}$$

Radiated power also can be gotten using the coefficients $\alpha(x, \omega)$ and $\beta(x, \omega)$ by the following equation:

$$Ps_2 = -\frac{1}{2} \int_{x_1}^{x_2} [\alpha(x, \omega) + \beta(x, \omega)] \mathrm{Re}[V(x)I^*(x)]dx \tag{21}$$

where $E(\omega)$ and $H^*(\omega)$ are the electric and magnetic field vectors on the planes surrounding the discontinuities lines structures except the bottom plane made of perfect constant and the incident, transmitted planes (Fig. 3).

The parameters along step microstrip line extracted by the generalized transmission line equations and the normal transmission line equations were given in Figs. 4 and 5 and the above-mentioned powers of the step line about frequency were shown in Figs. 6 and 7. All the data for Figs. 4 and 5 were calculated under the

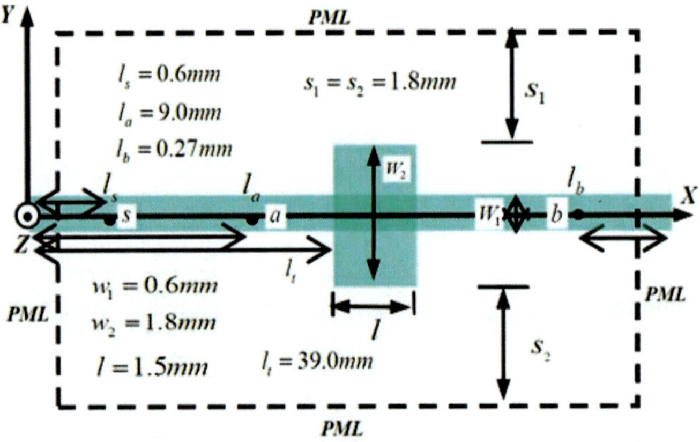

(a) Top view

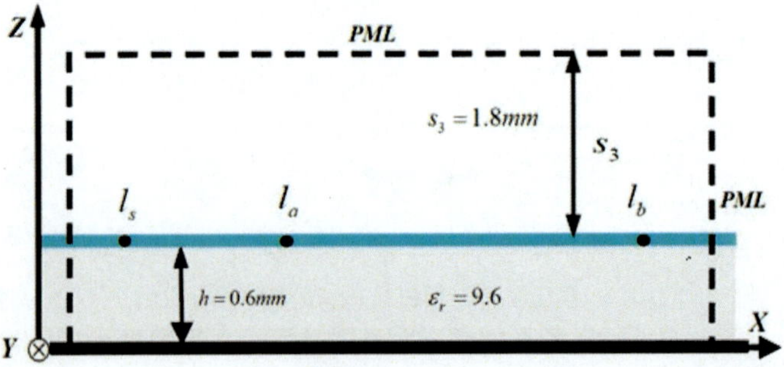

(b) Side view

Fig. 3 FDTD calculation domain for a step microstrip model

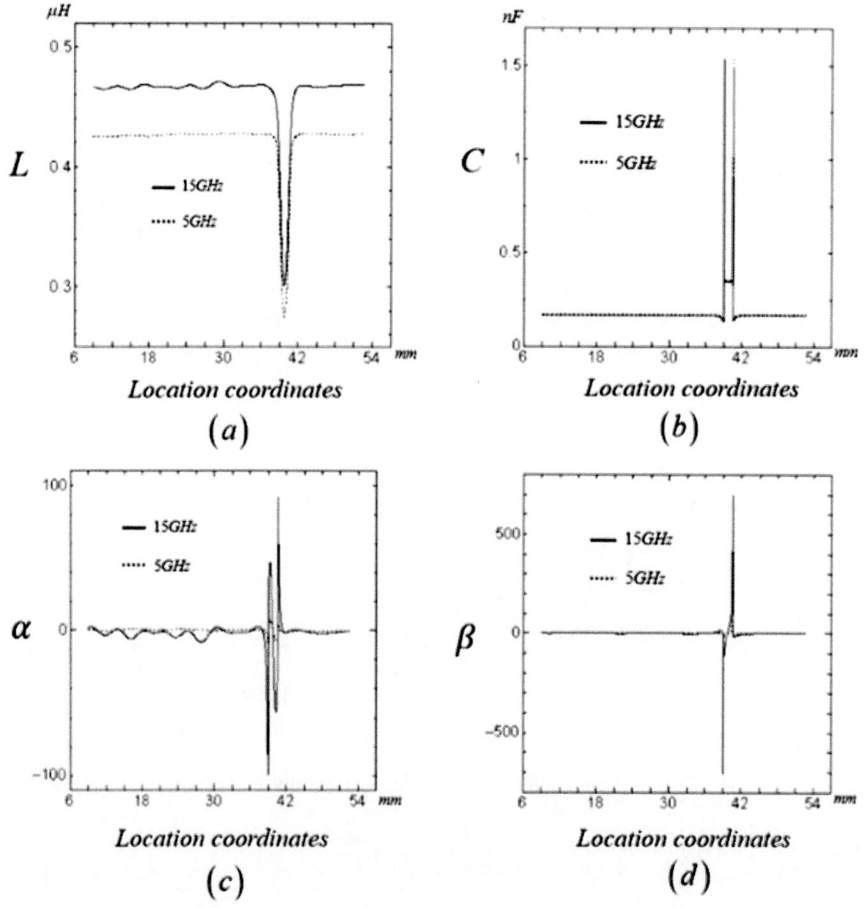

Fig. 4 *L*, *C*, *α* and ***β*** on the step strip line

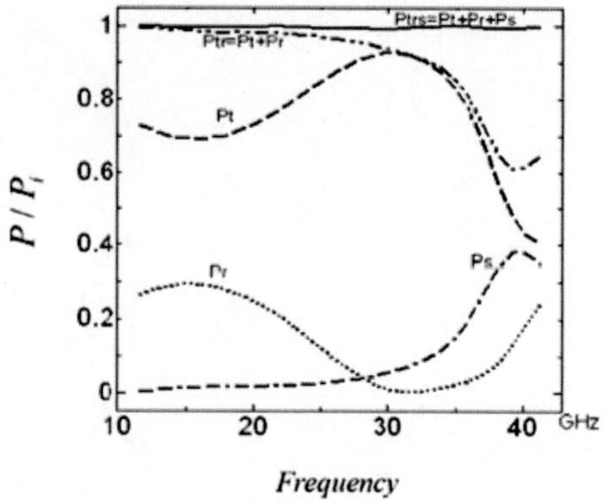

Fig. 5 *L*, *C*, *R* and *G* on the step strip line

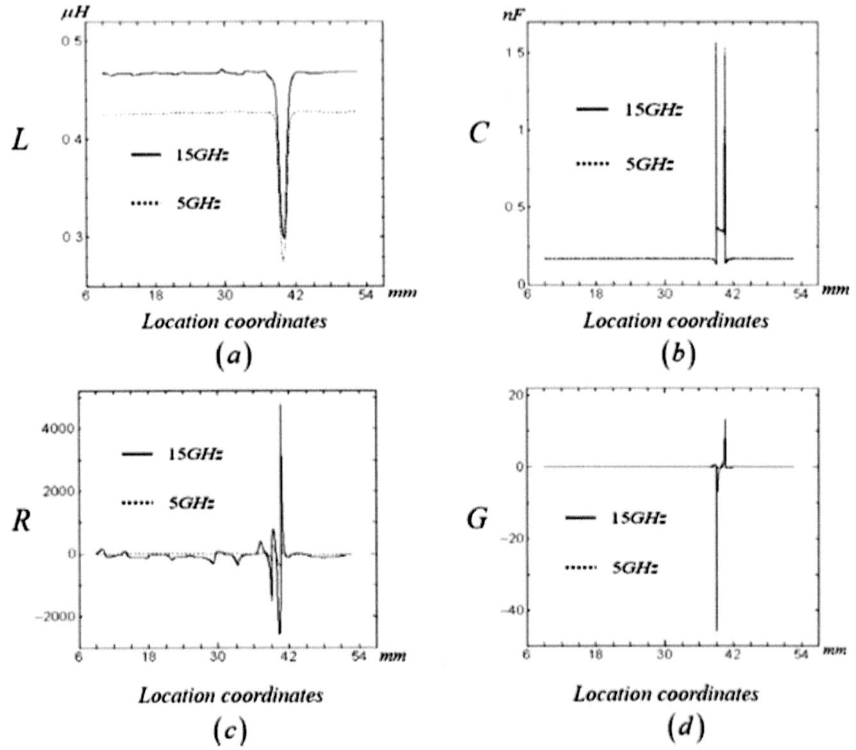

Fig. 6 Reflected power, transmitted power and radiated power of the step line by the FDTD

Fig. 7 Comparison of radiated power of the step line calculated by between the FDTD and the uniform transmission line equations model

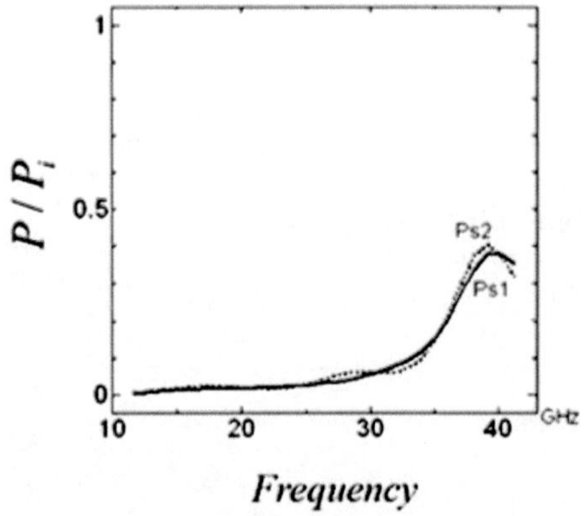

frequency of 5.0 and 15.0 GHz. We can see that the electromagnetic-wave circuit parameters which represent the local radiation and higher-order mode effect in the discontinuities segments have been extracted and it is confirmed from the results of the radiated power of the step line calculated by the FDTD coincided with by the uniform transmission line equations model.

4 Conclusion

In this paper, the problem of the discontinuities lines such as step microstrip line has been analyzed by the FDTD, The other discontinuities' structures can be analyzed with this method, such as Bend, T-junction, blench, Cross-junction and Cross-over. But we can see the parameters such as L, α and R became unstable when the frequency turned higher such as 15 GHz in this structure. It must be investigated and improved in the future.

References

1. Zhang XL, Mei KK (1988) Time-domain finite-difference approach to the calculation of the frequency-dependent characteristics of microstrip discontinuities. IEEE Trans Microw Theory Tech 36(12):1775–1787
2. Zhang XL, Fang JY, Mei KK, Liu YW (1988) Calculation of the dispersive character-istics of microstrips by the time-domain finite difference method. IEEE Trans Microw Theory Tech 36(2):263–267
3. Robertson I et al (2015) Rfic and mmic design and technology. Issues Des Technol Teach 11(4):92–93
4. Yan W, Zhao Y, Wang E (2012) Investigation and reduction on conducted electro-magnetic interference noise mechanism for complex power electronics systems. Proc Chin Soc Electr Eng 32(30):156–162
5. Kunz KS, Luebbers RJ (1993) The finite difference time domain method for Electromagnetics. CRC Press
6. Farahat N, Mittra R (2002) Extraction of equivalent-circuit parameters of interconnection lines with the use of the finite-difference time-domain method. Microwave Opt Technol Lett 34(1):59–61
7. Zhong X, Liu Y, Mei KK (2002) A new time-domain method based on the general transmis sion line equations. Microwave Opt Technol Lett 32(1):46–51
8. Liu Y, Mei KK (2002) Generalized transmission line equations. In: Handbook of small animal practice, vol 3, pp 1147–1156
9. Yee KS (1966) Numerical solution of initial boundary value problem involving maxwell's equations in isotropic media. IEEE Trans Antennas Propag AP-14(5):302–307
10. Hong JS, Liu YW, Wang BZ (2002) 2D generalized transmission-line equations for a microstrip cross. Microwave Opt Technol Lett 35(5):378–380
11. Hong JS, Wang BZ, Zhang Y (2004) Analysis of a double folded stub microstrip structure using generalized transmission line equations method. J Electromagn Waves Appl 18(9):1235–1247

Design of DMD Projection Lens for Structured Light 3D Measurement

Tengbiao Song, Lei Zhang, Yan Shi, Hongyao Liang, Haibin Niu, and Shangzhong Jin

Abstract To obtain 3D information of an object, the object must be measured. As a non-contact and active measurement method, structured light-based 3D measurement technology has a wide range of applications. The image quality of the projection lens determines whether the projection grating can be clearly projected on the detection object in the 3D measurement, which directly affects the detection effect. In this paper, a DMD projection imaging lens suitable for 3D measurement of structured light is designed. The effective focal length is 33.9 mm, F/# is 2.3, and the total length of the optical system is 120 mm. At a lens resolution of 93 lp/mm, the MTF values of all fields of view are greater than 0.5, the maximum distortion of the full field of view is less than 1%, the field curvature is less than 0.05 mm, and the imaging quality is good enough for 3D measurement.

Keywords 3D measurement · Projection lens · DMD · Optical design

1 Introduction

Optical 3D measurement is widely used in many fields such as industrial automatic detection, product quality control, reverse engineering, biomedicine, virtual reality, cultural relics reproduction, human body measurement and so on [1]. Optical 3D measurement technology is generally divided into two categories: passive 3D measurement and active 3D measurement according to different imaging illumination methods [2, 3].

T. Song · Y. Shi (✉) · H. Liang · H. Niu · S. Jin
College of Optical and Electronic Engineering, China Jiliang University, Hangzhou, Zhejiang, China
e-mail: Shiyan@cjlu.edu.cn

L. Zhang
Hangzhou HPWINNER OPTO Corporation, Hangzhou, Zhejiang, China

Y. Peng and X. Dong (eds.), *Proceedings of 2019 International Conference on Optoelectronics and Measurement*, Lecture Notes in Electrical Engineering 726,
https://doi.org/10.1007/978-981-33-4110-4_5

Active 3D measurement technology uses different projection devices to project different types of structured light to the measured object, and captures a structured light image that is deformed by modulation on the surface of the measured object. Calculate the 3D shape data of the measured object [4–6]. Of all the projection methods, the earliest, easiest and most used projection method is the slide projector projection [7].

There are three main types of structured light: point structured light, line structured light, and area structured light. Point structured light is a light spot formed by projecting a light beam onto the surface of the object to be measured. The light source is composed of a laser with high brightness and good directivity. Line structured light is a sheet of light that is projected onto the surface of the object to be measured. The surface forms a bright crossing line; the surface structured light projects a 3D figure onto the surface of the object to be measured, and forms a modulation pattern on the surface of the object [8].

In recent years, with the rapid development of DMD (digital micromirror device) technology, DMD-based surface structured light projection has become more and more widely used [9]. This paper designs a DMD projection lens with simple structure and high definition, which is suitable for 3D measurement system of structured light.

2 Design Ideas

2.1 Design Specifications

The projection distance is 450 mm, the projection size D is 6.7 inch (1 inch = 2.54 cm), the display chip size is 0.47 inch, and the entire system is designed backwards, and the magnification is [10]:

$$m = s/d = 0.07 \tag{1}$$

Lens focal length f is:

$$f = l/(1/m - 1) = 33.9 \tag{2}$$

The field of view of the projection lens is:

$$2\omega = 2\arctan(d/2l) = 20^\circ \tag{3}$$

In the formula: m is the magnification; s is the size of the display chip; f is the focal length of the projection lens; d is the size of the projection screen; l is the projection distance.

The chip size used in the design is 0.47 inch (11.94 mm), with 1920 × 1080 pixels, and the pixel size is 0.0054 mm × 0.0054 mm. In order to better match the

Table 1 Lens design parameters

Focal length (f/mm)	Field of view 2ω/(°)	F/#	Relative distortion	MTF (@93 lp/mm)	Chip size (Inch)
33.9	20	2.3	≤1%	≥0.5	0.47

resolution of the projection lens to the resolution of the chip, it must meet:

$$r = 1/2\mu \tag{4}$$

where: r is the resolution of the projection lens; μ is the chip pixel size. It can be seen that the characteristic frequency of the lens is 93 lp/mm. The specific parameters of the lens are shown in Table 1.

2.2 *Initial Structure Selection*

There are generally two methods for determining the initial structure of an optical system: one is to solve the initial structure parameters using the PW method; the other is to find the initial structure with similar technical indicators from existing structures in patents/documents and other materials. Because the first method has a large amount of calculation, this paper chooses the second method for design. According to the working characteristics of DMD, its projection lens is required to be telecentric to achieve high light efficiency and uniformity, that is, the main light of the DMD full field of view is perpendicular to the DMD surface [11]. The initial structure of the selected lens is shown in Fig. 1. The lens consists of 8 lenses with an effective focal length of 9 mm, an F/# of 2.5, a field angle of 76°, and a display chip size of 0.55 inch. Big gap.

2.3 *Optimized Design*

When designing the projection system, the wavelengths are 454 nm, 464 nm (dominant wavelength), and 474 nm. According to the design requirements, the focal length of the optical system is scaled so that the focal length is 33.9 mm, the F/# is adjusted to 2.3, and the curvature of each lens is set as a variable. The projection lens is a telecentric structure, and the telecentric degrees of the lens must be carefully controlled. In the automatic design optimization, the telecentric degrees of the system is con-trolled by the local incident angle so that it is 0°. In consideration of actual processing, to prevent the center of the concave lens from being too thin and the edge of the convex lens from being too thin, the boundary conditions are added to the automatically optimized structural limit so that the center thickness of the lens

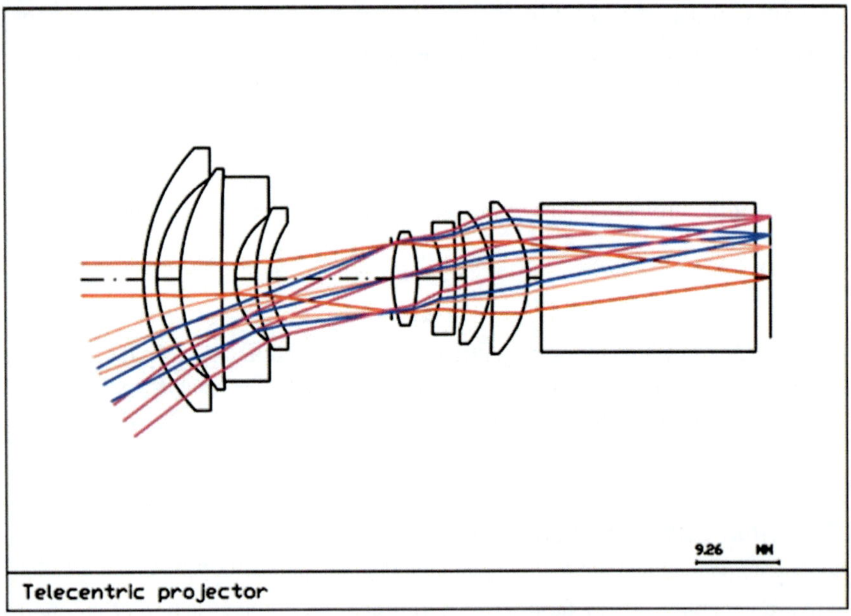

Fig. 1 Layout of initial structure of whole system

is greater than 0.85 mm and the edge thickness is greater than 0.65 mm. The glass in the initial structure was replaced with CDGM Environmentally Friendly Glass, and the initial structure was optimized. In the special control, the limit focal length is 33.9 mm, the total limit length is less than 120 mm, and the limit distortion is less than 1%. Through modification of constraints and weights, through automatic optimization and global optimization, and taking into account the aberration balance, the DMD projection lens that meets the design requirements is achieved.

3 Image Quality Analysis

After optimization, a grating projection lens with an effective focal length of 33.9 mm, an F/# of 2.3, a field of view of 20°, and a total system length of 120 mm is obtained. The layout of the optical path system is shown in Fig. 2.

The MTF(Modulation transfer function) curve, distortion and astigmatism curve, and dot map of this lens are shown in Figs. 3, 4 and 5. As can be seen from the figure, the MTF value of each field of view at 93 lp/mm is greater than 0.5. The transfer function curves of all fields of view are relatively concentrated and smooth, and the lens sharpness and contrast are relatively good. It can be considered that the imaging quality of the system is better. The maximum distortion of the lens is less than 1%,

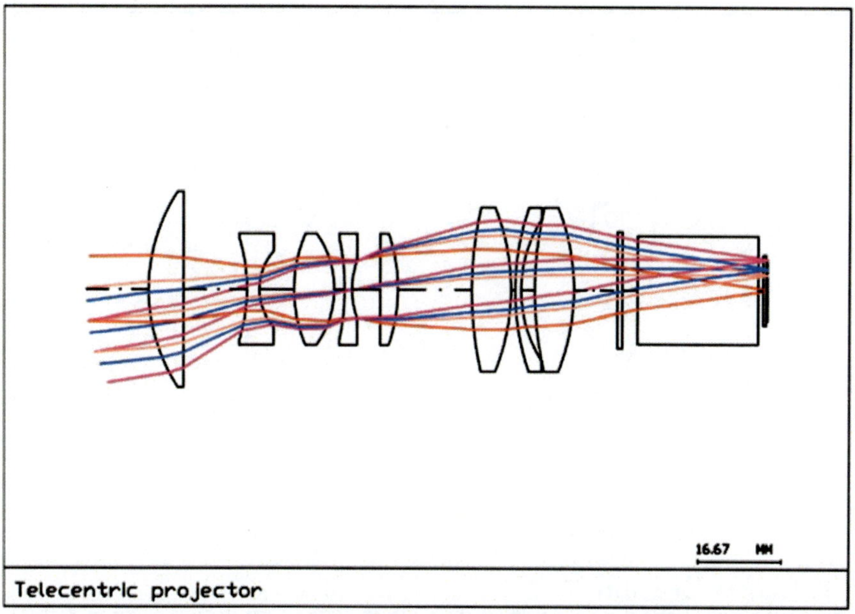

Fig. 2 Layout of whole system optical path after optimization

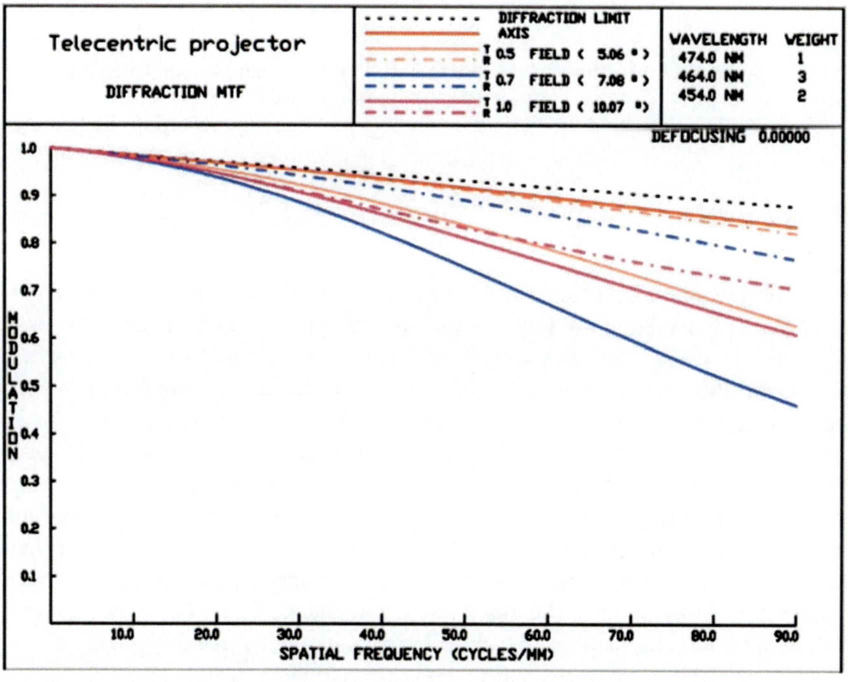

Fig. 3 Imaging quality obtained with the conventional optimization method: MTF

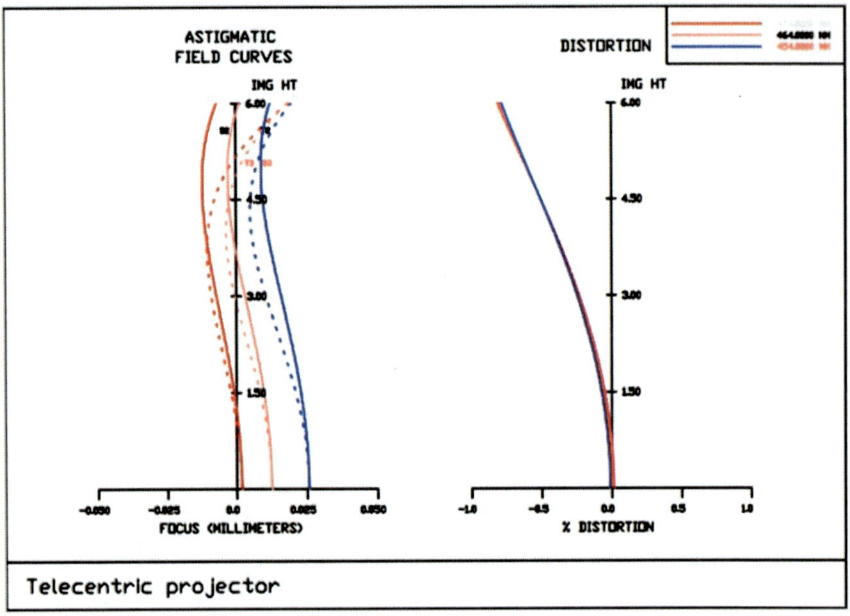

Fig. 4 Imaging quality obtained with the conventional optimization method: Field curvature and distortion curve

and the maximum field curvature is less than 0.05 mm. It can be seen from the figure that the aberrations of each field of view are better controlled.

4 Conclusion

This paper designs a DMD telecentric projection lens with a working band of 0.454–0.474 μm, a focal length of 33.9116 mm, an F/# of 2.3, and a full field of view of 20°. The lens design process first selects the appropriate initial structure, and uses various specific constraints in the optical software to restrict the basic parameters and the external dimensions of the lens. Finally, a grating projection lens suitable for 3D structured light measurement is obtained. This lens consists of 8 lenses. All lenses are spherical lenses without significantly reducing the performance of the projection lens, which greatly reduces the production difficulty and cost. The design results show that the maximum distortion of the lens is less than 1%, the maximum field curvature is less than 0.05 mm, the MTF value is higher than 0.5 when the spatial frequency is 93 lp/mm, and the maximum CRA value is 0.62°. From the results of various aberration evaluations, the optical system has better imaging quality.

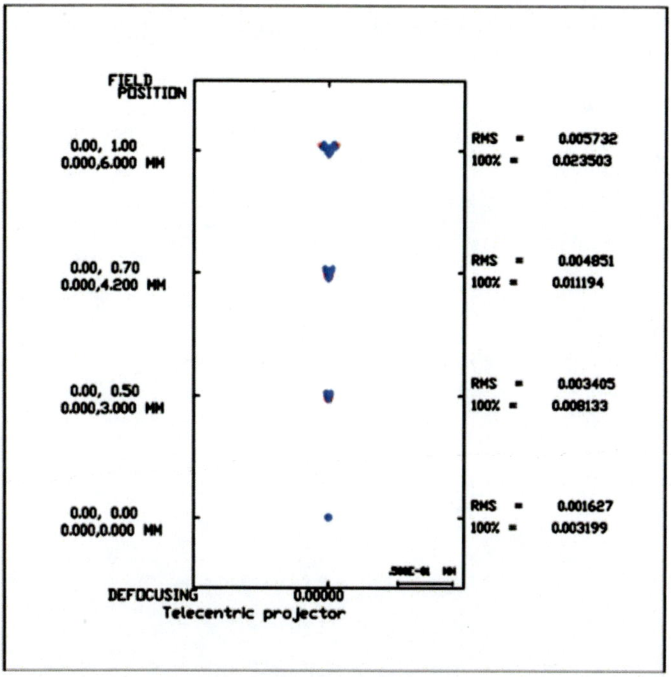

Fig. 5 Imaging quality obtained with the conventional optimization method: Spot diagram

Acknowledgments Funding. Major Scientific Research Project of Zhejiang Lab (2019DE0KF01).

References

1. Wang J, Bao HM, Wei ZH, He X, Hao ZH (2005) Optical 3D information measuring technology. Space Opt Dept Changchun Inst Opt Fine Mech Phys 02:32–36
2. Chen F, Brown GM, Song M (2000) Overview of three-dimensional shape measurement using optical methods. Opt Eng 39(1):10–22
3. Bkais F (2004) Review of 20 years of range sensor development. J Electron Imaging 13(1):231–243
4. Li ZW (2009) Research on structured light 3D measuring technology and system based on digital fringe projection. Huazhong Univ Sci Technol
5. Milroy MJ, Weir DJ, Bradley C, Vickers GW (1996) Reverse engineering employing a 3D laser scanner: a case study. Int J Adv Manuf Technol 12(2):111–121
6. Zhou WS, Su XY (1994) A direct mapping algorithm for phase-measuring profilometry. J Mod Opt 41(1):89–94
7. Hao YD, Zhao Y, Li DC (1998) Review of optical profilometry based on light pattern projection. Opt Technol 65:58–61
8. Liu XY (2019) Research on application of surface structured light in 3D measurement. Aeronaut Manuf Technol 62(10):83–87

9. Chen T, Lai JJ, Zhao Y (2006) Application of DMD in 3-D profilometry using grating projection. Infrared 09:24–28
10. Li XT, Cen ZF (2007) Geometrical optics, aberrations and optical design. Zhejiang University Press, Hangzhou
11. Zhong Y (2016) Optical design of the laser projection system based on DLP technology. Changchun Univ Sci Technol

Transfer Methods of CVD-Grown Two-Dimensional MoS_2: A Brief Review

Xinyu Song, Yinqin Ye, and Yingquan Peng

Abstract The thickness of two-dimensional (2D) layered materials is only a few hundred nanometers (nm), which has a unique nanoscale planar structure. Molybdenum disulfide (MoS_2) is a typical semiconductor two-dimensional atomic crystal material, which has excellent physical and electrical properties and so shows great potential in nano-electronic and optical applications. MoS_2 semiconductor has a direct band gap of 1.90 eV (single layer configuration) and an indirect band gap of 1.29 eV (multi-layer configuration). With the weak van der Waal's forces between the layers, the monolayer MoS_2 shows more excellent properties in optoelectronic devices than graphene with zero band gap. Transferring molybdenum disulfide film from growth substrate to target substrate is a key problem in practical application. From the original micromechanical stripping to the most widely used wet etching method, more methods have been discovered, in order to improve the transfer quality of molybdenum disulfide films. This review focuses on discussing about ultrasonic bubbling transfer and surface energy assisted transfer methods, which improve the transfer quality of MoS_2 thin film.

Keywords Transfer methods · MoS_2 thin film · Ultrasonic bubbling transfer · Surface energy assisted transfer

1 Introduction

Two-dimensional (2D) materials have attracted substantial attention due to their abundant active sites and a unique nanoscale planar structure. Transition metal dichalcogenides (TMDs) [1], especially MoS_2, as one of the 2D materials were widely studied. In the synthesis and fabrication of 2D materials, especially in large

X. Song · Y. Ye · Y. Peng (✉)
College of Optical and Electronic Technology, China Jiliang University, Hangzhou, Zhejiang 310018, People's Republic of China
e-mail: yqpeng@cjlu.edu.cn

Y. Peng and X. Dong (eds.), *Proceedings of 2019 International Conference on Optoelectronics and Measurement*, Lecture Notes in Electrical Engineering 726,
https://doi.org/10.1007/978-981-33-4110-4_6

area chemical vapor deposition (CVD) fabrication [2], it is important to choose a perfect initial substrate for the growth of two-dimensional materials. As we know, MoS_2 can be synthesized by CVD on different substrates with atomically smooth surfaces such as SiO_2 on Si, sapphire, and mica [3]. It is reported that recent achievements have been made on metalorganic chemical vapor deposition (MOCVD), low-pressure chemical vapor deposition (LPCVD) and atmospheric pressure chemical vapor deposition (APCVD) about synthesis methods of 2D materials MoS_2 [4]. The synthesis methods of molybdenum disulfide include micromechanical exfoliation, liquid phase exfoliation, physical vapor deposition, solution chemical process and CVD. Transferring MoS_2 thin film to a suitable substrate is the key to the subsequent preparation of modern optoelectronic devices. These require the growth of MoS_2 thin films to be uniform and continuous, and can be transferred to arbitrary substrate with high fidelity. In this paper, we provide a brief literature review of the latest progress in the transfer methods of MoS_2 thin films. The paper focuses only on the transfer methods rather than progress in the synthesis of MoS_2. The transfer of the synthesized MoS_2 thin film to arbitrary substrate is an important step for optoelectronic devices, especially for flexible devices.

2 Transfer Methods of MoS_2

2.1 Micromechanical Exfoliation

Micromechanical exfoliation is the simplest and most efficient method to prepare two-dimensional nanomaterials [5–7]. Similar to the micromechanical peeling of graphene, MoS_2 films can be obtained by the simplest manual method without the need of complex and expensive instruments. MoS_2 thin film is synthesized by CVD with MoO_3 and sulfur powder, and then it was torn off by tape and transferred to the substrate to be transferred. After the tape is released several times, due to van der Waals force, some material of MoS_2 remain on the substrate instead of the tape. This process may produce MoS_2 thin film with random shape, size and number of layers. Magda et al. obtained mono MoS_2 layers with a transverse dimension of several hundred microns by improving the adhesion between MoS_2 and substrate, as shown in Fig. 1 [5].

As a result, the MoS_2 thin film fabricated by this method have a few defects and a perfect crystal structure. Moreover, it has good optoelectronic properties, such as a high mobility. But this method also causes several great shortcomings, such as small size and poor repeatability of molybdenum disulfide samples, so the method from micromechanical exfoliation of transfer MoS_2 thin film method is only suitable for laboratory research, not large-scale production.

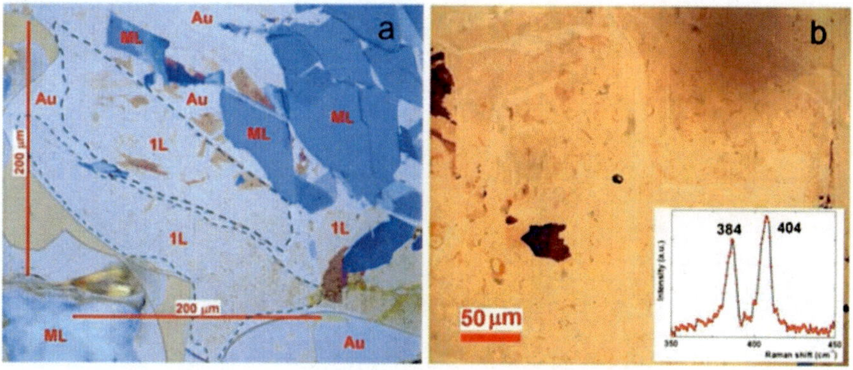

Fig. 1 Large area MoS_2 thin film exfoliated with chemically enhanced adhesion to the substrate. Reprinted with permission from Ref. [5]. Copyright © 2015, Springer Nature. All rights reserved

2.2 Wet Etching Transfer

Suk et al. have realized the transfer of CVD-grown monolayer graphene onto arbitrary substrate [8], and the transfer method of MoS_2 thin films imitates this method. Wet etching transfer method is recognized as the most popular and popular technology at present.

In this wet etching method, a drop of PMMA solution (6 mg/mL) is spin-coated on MoS_2 thin film at 3000-4000 rpm for 30 s and heated at 180 °C for 90 s. The PMMA/MoS_2/substrate stack is then immersed in a 2 mol/L KOH solution at 20–90 °C until the MoS_2 thin film falls off [8–11]. However, it is found that this method is difficult to make the thin film fall off naturally. It is necessary to use knife or tweezers to treat the edge of MoS_2 thin film in combination with the above-mentioned micromechanical exfoliation method to promote the falling off. After rinsing MoS_2 thin film with deionized water, the target substrate is placed in deionized water with a tilting angle of 30–45° under the floating MoS_2 thin film and heated at 80 °C for 30 min on vacuum drying oven. PMMA is removed in acetone solution for 3–6 h, then washed under deionized water, and heated at 100 °C for about 10 min on vacuum drying oven. In different literatures, in order to improve the quality of film transfer, some materials are replaced, such as PS instead of PMMA [12, 21], NaOH [12, 13] or HF [14, 17] instead of KOH solution, chloroform instead of acetone solution [15]. Before the transfer, the substrate to be transferred needs to be cleaned with acetone, ethanol and deionized water.

On the basis of wet transfer, a layer of Au or Cu can be deposited on MoS_2 thin film by electron beam deposition, and the MoS_2/Au sample was spin coated with PMMA solution and heated at 170 °C for 10 min. The sample was then soaked in Au etchant (tape TFA) for the removal of Au [16]. The later transfer method is consistent with the above method. Mlack et al. use tape and PMMA bonding method to transfer monolayer tungsten disulfide [18]. This method can also be applied to MoS_2, which makes the transferred film have less wrinkles and cracks. This method is consistent

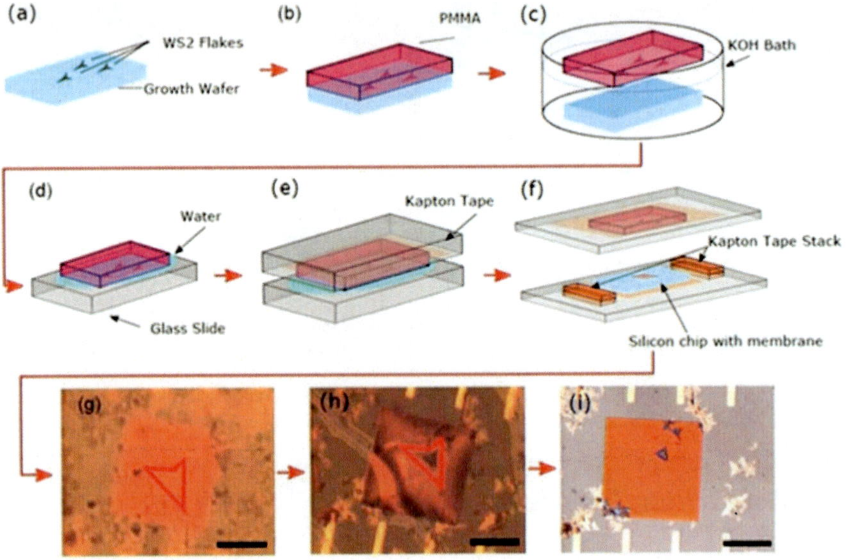

Fig. 2 Transfer images of two-dimensional WS_2. Reprinted with permission from Ref. [18].

with the above method before the film is pulled out of water with glass slide, and the glass slide is fixed on the suspended PMMA with special tape. The slide with PMMA is aligned with the target substrate fixed on the separate slide, and two tapes are used to stick the two slides together, as shown in Fig. 2.

2.3 *Ultrasonic Bubbling Transfer*

Figure 3 shows that molybdenum disulfide is transferred by ultrasonic bubbling [20], and PMMA solution is spin-coated on the MoS_2 thin film. In order to maintain the integrity of the film, it is necessary to spin coat the film to be thicker than in the above method, spin coating the thickness of 0.3 um at a low speed of 1000 rpm. The MoS_2 thin film needs to be annealed at 180 °C for 15 min to completely remove the residual PMMA solvent during spin coating. This annealing method needs to be used after spin coating, which can eliminate the influence of residual PMMA solution on the experiment. The PMMA/MoS_2/substrate is put into the breaker, and the breaker is immersed in ultrasonic cleaning device. The edge of the PMMA/MoS_2 thin film falls out from the substrate to the deionized water surface [19–22]. When ultrasonic wave vibrates under negative pressure, a lot of cavitation bubbles are produced, as shown in Fig. 3a. When the pressure of ultrasonic vibration suddenly changes from negative to positive, these bubbles quickly expand into larger bubbles. In a short time, these bubbles release an enormous amount of energy while compressed

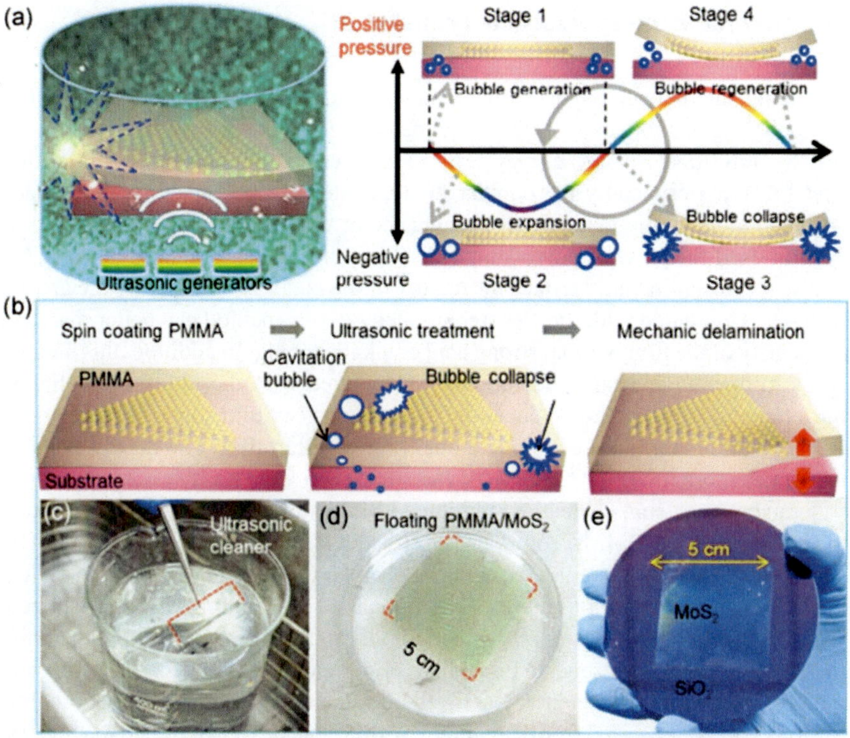

Fig. 3 Illustration of ultrasonic bubbling transfer of MoS_2. Reprinted with permission from Ref. [20]. Copyright Springer Nature BV. All rights reserved

during the positive pressure period, which produces a great force in the interface between the PMMA/MoS_2 thin film and substrate. In multiple bubbling cycles, the PMMA/MoS_2 was stably separated from substrate by bubbling inducing force. The interface between PMMA and MoS2 is very stable in ultrasonic process, because the adhesion between them is very strong, and ultrasonic vibration is not enough to separate them. The delaminated PMMA/MoS_2 thin film can be transferred onto arbitrary target substrate [20]. According to this hydrophobicity, a method of surface energy assisted transfer on transfer MoS_2 thin film appears later.

Compared with the wet etching method, this method can better maintain the integrity of the film without strong acid or alkali etching. In addition, this method is easy to operate and takes a short time. The whole layering process only uses water, and does not involve chemical corrosives or harmful pollutants.

2.4 Surface Energy Assisted Transfer

The relationship between hydrophilicity and hydrophobicity derived from the ultrasonic bubbling method above extends the method of surface energy assisted transfer of MoS_2 thin films. Surface energy assisted transfer can be transferred to arbitrary target substrate, especially the popular flexible substrate, and then it can be made into flexible electronic devices. At present, the acid wet chemical etching process which is used to transfer molybdenum disulfide film in large area is not suitable because it can significantly reduce the quality of the material and destroy the growth substrates. Compared with the wet etching method, this method can better maintain the integrity of the film without strong acid or alkali etching. In addition, this method is easy to operate and takes a short time. Compared with the ultrasonic bubbling transfer method, this method reduces the micro damage of ultrasonic vibration to the film, and can obtain a more complete quality MoS_2 thin film. The disadvantage is that it takes more time.

Gurarslan et al. spin coated PS solution on MoS_2 thin film [21]. The PS solution replaces the previous PMMA solution. The effect of PS solution on spin coating a few layers of film (especially single layer) is better than that of PMMA solution. When the number of spin coating film layers is less than 10, PS solution can be considered. Cover PS with a drop of deionized water, and then treat the edge of PS/MoS_2 thin film with a knife or tweezers, and deionized water can enter between PS/MoS_2 and sapphire. When the surface energy assisted between them reaches a certain degree, PS/MoS_2 will be separated from sapphire. PS/MoS_2 can be transferred to the SiO_2/Si substrate, baked at 80° for 1 h to remove the water residue and baked at 150° for 30 min to remove the possible wrinkles [21] and finally washed with toluene solution for several times to remove PS. Optoelectronic devices can be fabricated after drying. Figure 4 shows the surface energy assisted transfer process.

Figure 5 shows that the substrate is dealed with diluted hydrofluoric (HF:H_2O = 1:10) before the synthesis of MoS_2 based on the surface-energy-assisted transfer method to make the substrate super hydrophilic. SiO_2/Si is spin coated with PMMA solution, and then slowly immersed in deionized water. Because of the difference in surface energy assisted, the deionized water immediately penetrates PMMA/MoS_2 and the substrate, causing PMMA/MoS_2 to float on the deionized water. This method is suitable for large area wafer-scale MoS_2 transfer to flexible substrate and fabrication of flexible optoelectronic devices.

In order to reduce the wrinkles and cracks during the transfer of MoS_2 thin film, it is found that deionized water and alcohol 1:1 are mixed instead of deionized water on the basis of surface energy assisted transfer, and the wrinkles and cracks are greatly reduced as shown in Fig. 6. Wrinkle area ratio from 4.11 to 0.76% [22]. Figure 6c and f can more intuitively show the comparison between the two solutions for MoS_2 thin film transfer.

Figure 7c and d show the contact angles of deionized water and mixed liquid (deionized water: alcohol = 1:1) on HfO_2/Si substrate, respectively [22]. When mixed with alcohol, the contact angle decreases obviously, which may be due to reduction

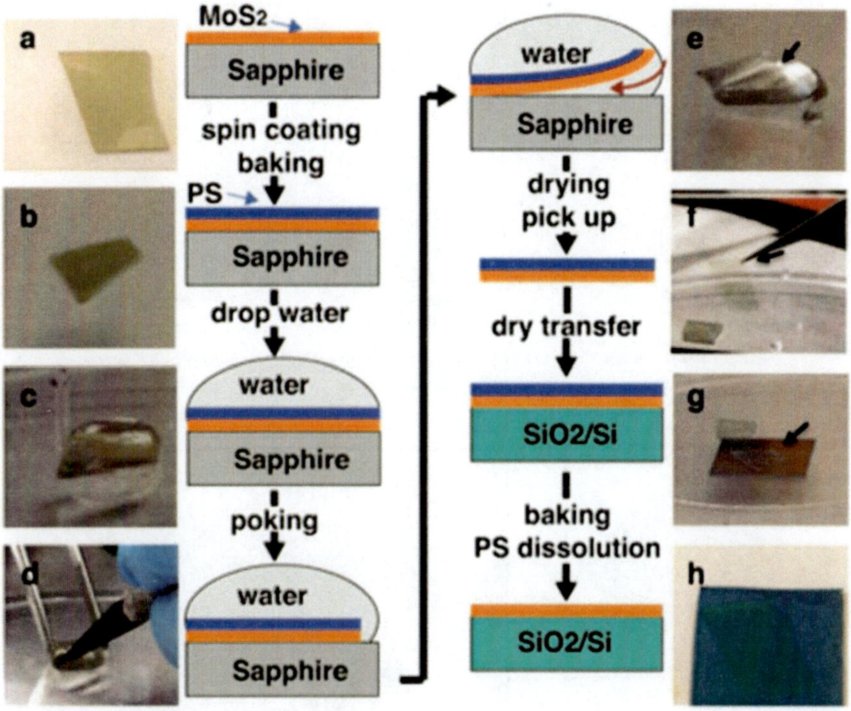

Fig. 4 Illustration of the surface energy assisted transfer process. Reprinted with permission from [21].

of surface tension γ_{sl} by the surfactant action of alcohol in water. In addition, metal oxide semiconductor is a kind of hydrophobic material. Mixed alcohols as surfactants may also reduce interfacial tension γ_{fl}. Considering three kinds of surface tension, mixing deionized water with ethanol will increase diffusion parameters and lead to wrinkle free MoS_2. In addition, the ratio of deionized water to ethanol depends on the surface tension of the substrate used. For HfO_2/Si substrates, the ratio can be between 1:1 and 3:1 [22]. However, when a high proportion of alcohol is used, it will have a negative effect. For the transfer ratio of other substrates, it is not mentioned that it is necessary to calculate the effect of three surface tensions on the substrate to be transferred and calculate the appropriate ratio.

3 Conclusions

In this paper, we have reviewed several transfer methods of MoS_2 in terms of micromechanical exfoliation, wet etching transfer, ultrasonic bubbling transfer and surface energy assisted transfer. Micromechanical exfoliation is simple and fast,

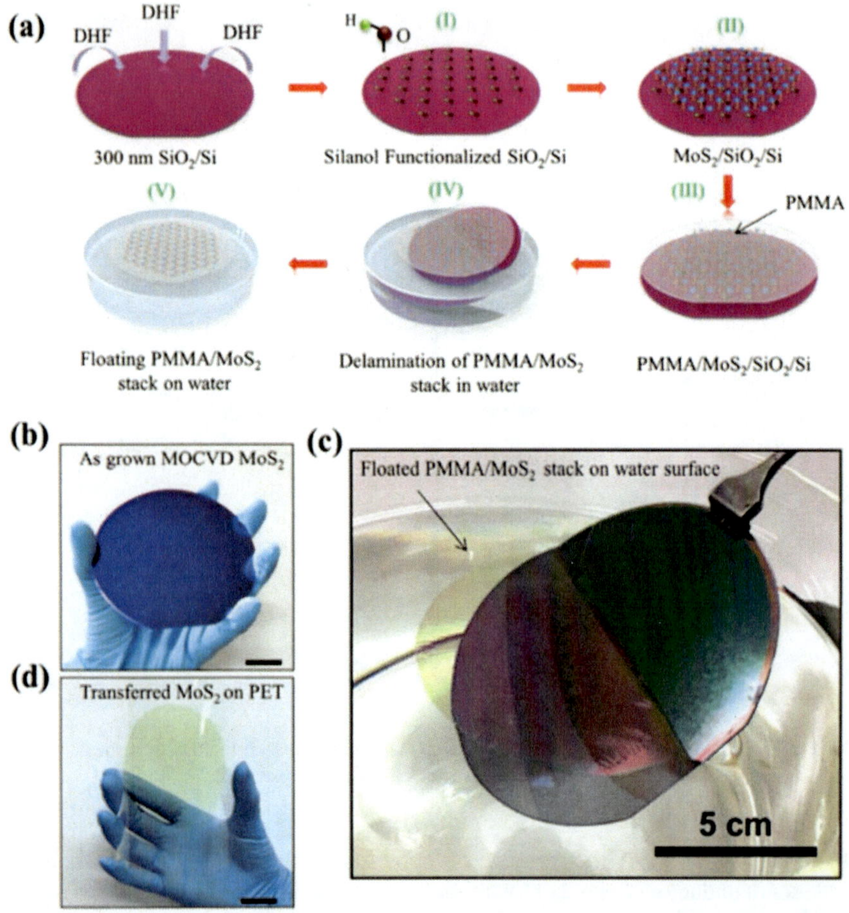

Fig. 5 **a** Surface functionalized transfer of molybdenum disulfide over a large area. **b** As grown MOCVD MoS_2.. **c** Transferred MoS_2 on PET. **d** Floated PMMA/MoS_2 stack on water surface. Reprinted with permission from Ref. [17], Advanced Functional Materials Publishing. Copyright the "Wiley Materials". All rights reserved

but the transfer size is uncontrollable. Wet etching transfer is the most popular and common method at present. In this review, we have put a special emphasis on the ultrasonic bubbling transfer and surface energy assisted transfer methods, because the films transferred by these two methods are of higher quality and repeatability, they can be transferred in a large area and can be applied to the hottest flexible substrates. Flexible optoelectronic devices can be fabricated by flexible substrates. With the increasing maturity of the flexible substrate process, the requirements for the transfer process will be higher. This high-quality, high-repeatability, large-scale transfer method will be suitable for large-scale industrial production.

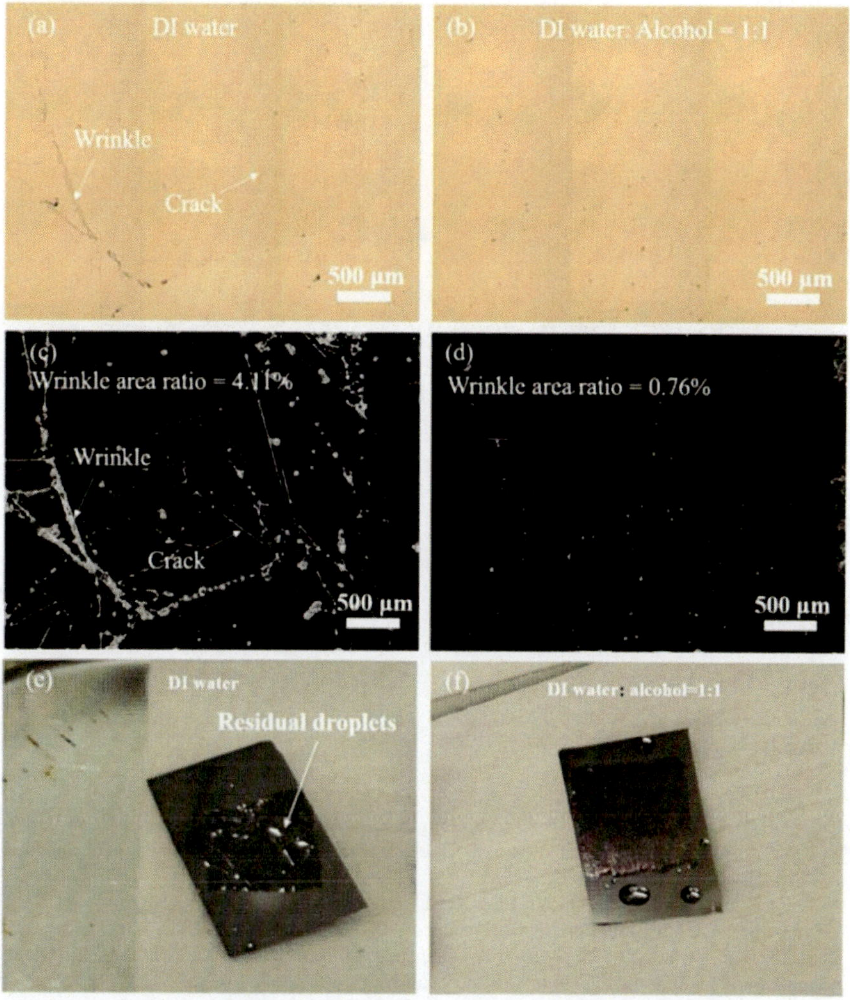

Fig. 6 Comparison of transfer results of two solutions for MoS_2 thin film. Reprinted with permission from Ref. [22], and © IOP Publishing. Reproduced with permission. All rights reserved

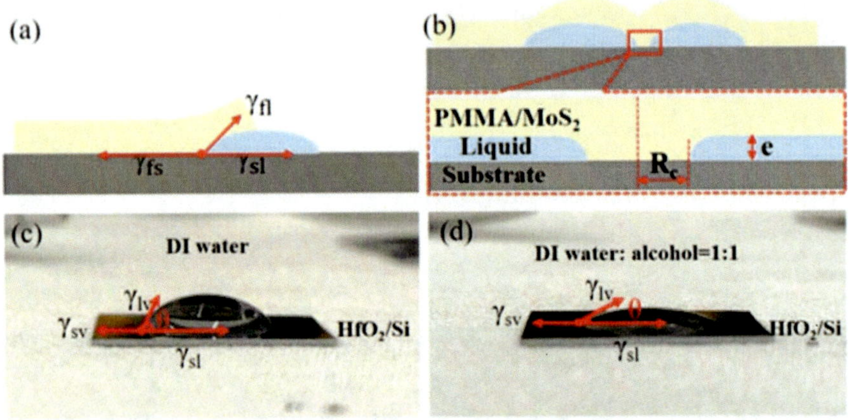

Fig. 7 Graphic vector representation of three surface tensions in two different solutions. Reprinted with permission from Ref. [22], and © IOP Publishing. Reproduced with permission. All rights reserved

Acknowledgements This work was supported by the Zhejiang Provincial Natural Science Foundation of China Grant No. LQ19F040003.

References

1. Kang SB, Kwon KC, Choi KS, Lee R, Hong K, Suh JM (2018) Transfer of ultrathin molybdenum disulfide and transparent nanomesh electrode onto silicon for efficient heterojunction solar cells. Nano Energy 50:649–658
2. Lee YH, Yu L, Wang H, Fang W, Ling X, Shi Y (2013) Synthesis and transfer of single-layer transition metal disulfides on diverse surfaces. ACS Publ 13(4):1852–1857
3. Lai S, Jeon J, Song YJ, Lee S (2016) Water-penetration-assisted mechanical transfer of large-scale molybdenum disulfide onto arbitrary substrates. RSC Adv 6(62):56747–57772
4. Xu ZQ, Zhang Y, Lin S, Zheng C, Zhong YL, Xia X (2015) Synthesis and transfer of large-area monolayer WS_2 crystals: moving toward the recyclable use of sapphire substrates. ACS Publ 9(6):6178–6187
5. Magda GZ, Pető J, Dobrik G, Hwang C, Biró LP (2015) Exfoliation of large-area transition metal chalcogenide single layers. Sci Rep 5(1):14714
6. Frindt RF (1996) Single crystals of MoS_2 several molecular layers thick. J Appl Phys 37(4):1928–1929
7. Yu H, Liao M, Zhao W, Liu G, Zhou XJ, Wei Z, Xu X (2017) Wafer-scale growth and transfer of highly-oriented monolayer MoS_2 continuous films. ACS Publ 11(12):12001–12007
8. Suk J, Kitt A, Magnuson CW, Hao Y, Ahmed S, An J (2011) Transfer of CVD-grown monolayer graphene onto arbitrary substrates. ACS Publ 5(9):6916–6924
9. Najmaei S, Liu Z, Zhou W, Zou X, Shi G, Lei S (2013) Vapour phase growth and grain boundary structure of molybdenum disulphide atomic layers. Nat Mater 12(8):751–759
10. Van AM, Huang PY, Chenet DA (2013) Grains and grain boundaries in highly crystalline monolayer molybdenum disulphide. Nat Mater 12(6):554–561
11. Elías AL, Perea-López N, Castro-Beltrán A (2013) Controlled synthesis and transfer of large-area WS_2 sheets: from single layer to few layers. ACS Publ 7(6):5235–5242

12. Kim JY, Kim DH, Kwon MK (2017) Controlled growth of large-area and high-quality molybdenum disulfide. Jap J Appl 5(6):110302.1–110302.4
13. Salvatore GA, Münzenrieder N, Barraud C, Petti L (2013) Fabrication and transfer of flexible few-layers MoS_2 thin film transistors to any arbitrary substrate. ACS Publ 7(10):8809–8815
14. Li H, Wu J, Huang X, Yin Z, Liu J, Zhang H (2014) A universal, rapid method for clean transfer of nanostructures onto various substrates. ACS Publ 8(7):6563–6570
15. Sun J, Li X, Guo W, Zhao M, Fan X, Dong Y, Xu C (2017) Synthesis methods of two-dimensional MoS_2: a brief review. Crystals 7(7):198
16. Shi J, Ma D, Han GF, Zhang Y, Ji Q, Gao T, Sun J (2014) Controllable growth and transfer of monolayer MoS_2 on Au Foils and its potential application in hydrogen evolution reaction. ACS Publ 8(10):10196–10204
17. Shinde SM, Das T, Hoang AT (2018) Surface-functionalization-mediated direct transfer of molybdenum disulfide for large-area flexible devices. Adv Function 28(13):1706231.1–1706231.11
18. Mlack JT, Das PM, Danda G, Chou YC, Naylor CH (2017) Transfer of monolayer TMD WS_2 and Raman study of substrate effects. Sci Rep 7(1):43037
19. Yu Y, Fong PW, Wang S, Surya C (2016) Fabrication of WS 2/GaN pn junction by Wafer-Scale WS 2 thin film transfer. Sci Rep 6:37833
20. Ma D, Shi J, Ji Q, Chen K, Yin J, Lin Y, Zhang Y, Liu M (2015) A universal etching-free transfer of MoS_2 films for applications in photodetectors. Nano Res 8(11):3662–3672
21. Yu SY, Su L, Yu Y, Suarez F, Yao S, Zhu Y (2014) Surface-energy-assisted perfect transfer of centimeter-scale monolayer and few-layer MoS_2 films onto arbitrary substrates a gurarslan. ACS Publ 8(11):11522–11528
22. Liu X, Huang K, Zhao M, Li F, Liu H (2019) A modified wrinkle-free MoS2 film transfer method for large area high mobility field-effect transistor. Nanotechnology 31(5):055707

Organic Near Infrared Photodiode Based on Tin Naphtalocyanine as Sensitive Layer

Xinyu Song and Wenli Lv

Abstract The organic photodiodes (OPDs) have the advantages of rich functional material, controllable light-sensitive wavelength, low cost, and can be prepared on a flexible substrate, and has an important application prospect in the aspects of image sensing, immune detection, optical communication. Near infrared (NIR) optical devices can be fabricated mainly by using narrow energy gap materials and organic near infrared dye doping. Metal phthalocyanine, metal naphtalocyanine/fullerene C_{60} planar heterojunctions (PHJs) were often used as the active layer of NIR OPD. In this paper, NIR OPD based on SnNc as near infrared photosensitive layer were studied, because of its excellent absorption efficiency in NIR region. Bi-layer SnNc/C_{60} PHJ based OPD exhibited a large dark current, resulting a low photosensitivity. To reduce the dark current of the device, CuPc was introduced as electron blocking layer in the device, and tri-layer CuPc/SnNc/C_{60} PHJ based OPD was studied. The device shows improved performance, such as a low dark current, a large photoresponsivity, photosensitivity, the external quantum efficiency and the specific detectivity. The improved performance of device can be attributed to the enhanced electron injection barrier at the interface of anode and CuPc electron blocking layer.

Keywords Organic photodiodes · Near infrared · Planar heterojunctions

1 Introduction

Near infrared (NIR) field involves many industrial applications, such as artificial optical communication [1], biomedical science, imaging system [2] and remote monitoring. The organic photodiodes (OPDs) have the advantages of rich functional material, controllable light-sensitive wavelength, low cost, and can be fabricated on

X. Song · W. Lv (✉)
College of Optical and Electronic Technology, China Jiliang University, Hangzhou, Zhejiang 310018, People's Republic of China
e-mail: lvwl@cjlu.edu.cn

Y. Peng and X. Dong (eds.), *Proceedings of 2019 International Conference on Optoelectronics and Measurement*, Lecture Notes in Electrical Engineering 726,
https://doi.org/10.1007/978-981-33-4110-4_7

flexible substrates, and has an important application prospect in the aspects of image sensing, immune detection, optical communication. Near infrared optical devices can be fabricated by doping narrow gap materials and organic near infrared dyes. Metal phthalocyanines such as lead phthalocyanine (PbPc) [3], tin phthalocyanine (SnPc) [4–7] and aluminium phthalocyanine chloride (AlPcCl) are mainly used organic narrow energy gap materials and near infrared absorbing materials for OPD, organic solar cells and photoresponsive organic field effect transistors.

In OPD devices, charge carriers are generated through the dissociation of photogenerated excitons in the region where the electric field is sufficiently strong. Donor (D)/acceptor (A) PHJ structure can effectively improve the photo-generated exciton dissociation efficiency, due to the high electric field in the D/A interface. Metal phthalocyanines/fullerene C_{60} planar heterojunctions (PHJs) were often used as the active layer of NIR OPD [8]. In this paper, NIR OPD based on SnNc as NIR photosensitive layer were studied. Bi-layer SnNc/C_{60} PHJ based OPD exhibited a large dark current, resulting a low photosensitivity. In order to reduce the dark current of the device, CuPc was introduced as electron blocking layer in the device, and tri-layer CuPc/SnNc/C_{60} PHJ based OPD was studied. The device shows improved performance, such as a low dark current, a large photoresponsivity, photosensitivity, the external quantum efficiency and the specific detectivity.

2 Experimental Section

CuPc was purchased from J&K Chemical Ltd., C_{60} from Luminescence Technology Co., Ltd., Taiwan, and BCP and SnNc from Jilin OLED Co. Ltd. China. 30Ω/square patterned ITO glass substrates were used for fabricating device, the substrates were cleaned by an Ultrasonic Cleaner in acetone, ethanol, and de-ionized water stepwise, and then blow-dried with high pressure N_2 gas and desiccated in a vacuum drying oven (60 °C). Device A and B with the organic active layer of "SnNc/C_{60}/BCP" and "CuPc/SnNc/C_{60}/BCP" (Fig. 1) respectively was fabricated by thermally evaporating the active materials on the ITO glass substrates in sequence. A 100 nm thick aluminum top electrode with an active area of about 1.5 mm^2 was deposited by shadow mask vacuum deposition. The deposition rate of the evaporated material is 1–2 nm min^{-1}, and the thickness of the film is controlled by the quartz crystal oscillator near the sample. This measurement data was measured with the incident light power density of 100 mW/cm^2 and the wavelength of 850 nm red laser as the light source, and the current-voltage (*I-V*) characteristics are tested by self-made measuring system.

3 Results and Discussion

Figure 2 shows the UV-Vis absorption spectra of CuPc, C_{60} and SnNc films on quartz glass, respectively. One can see that SnNc thin film exhibit strong absorption at the

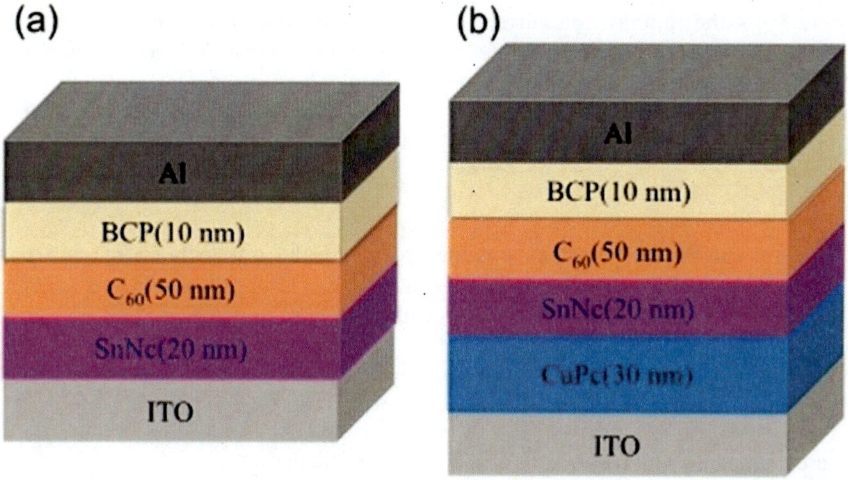

Fig. 1 The schematic structures of **a** device A and **b** device B

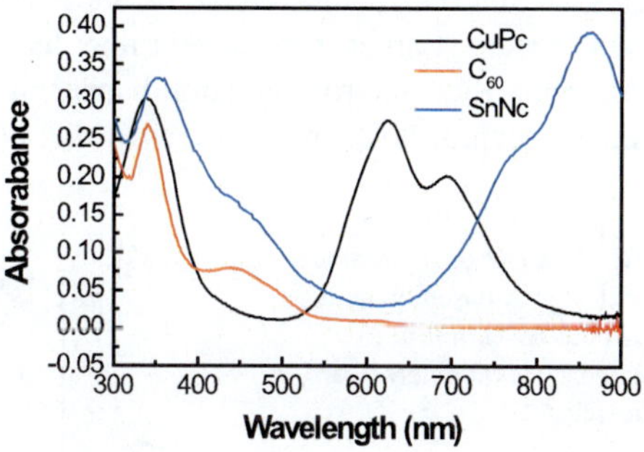

Fig. 2 Absorption spectra of CuPc, C_{60} and SnNc thin films on the quartz substrates

wavelength around 800–900 nm, while, CuPc and C_{60} exhibit negligible absorption at the same region. As a result, in the NIR region the absorption of SnNc/C_{60} and CuPc/SnNc/C_{60} PHJ films originates mainly from SnNc molecule.

The photosensitivity (P) and photoresponsivity (R) are two important performance parameters of photodiode. The ratio of the photo-current (I_{ph}) to the dark current (I_{dark}) is photosensitivity [9] and the ratio of photo-current to the incident optical power (P_{opt}) is photoresponsivity. The P and R of the device were calculated by the following relation.

$$P = \frac{I_{ph}}{I_{dark}} = \frac{I_{ill} - I_{dark}}{I_{dark}} \tag{1}$$

$$R = \frac{I_{ph}}{P_{opt}} = \frac{I_{ph}}{P_{in}A} \tag{2}$$

where I_{ill} is the illumination current, P_{in} is incident light power density and A is the effective area of device. Based on the photoresponsivity, the external quantum efficiency (*EQE*) can be calculated [10]:

$$EQE = \frac{hc}{q\lambda}R \tag{3}$$

where h is the Plank's constant, c is the speed of light, q is meta charge and λ is the wavelength of incident light. At reverse voltages larger than zero, the specific detectivity ($D*$) can be calculated by the following relation [11]:

$$D* = R/\sqrt{2qJ_{dark}} \tag{4}$$

where J_{dark} is the current density of the device in the dark.

I-V characteristics of Device A based on active layer at 850 nm laser in the dark and under NIR illumination as shown in Fig. 3a. When the reverse bias increases, the dark current increases rapidly, and reaches 90.83 uA at the reverse bias voltage of −5 V. The current is 278.07 uA under NIR illumination, which is ~3 times of the dark current. Larger dark current may lead to lower photosensitivity. As shown in Fig. 3b, a low electron injection barrier of 1 eV exists at the interface of ITO and SnNc acceptor layer, with the result that electrons be effectively injected from the

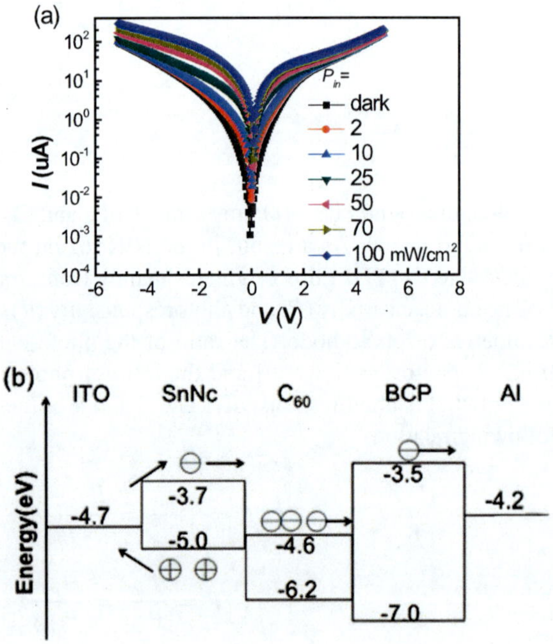

Fig. 3 **a** *I-V* characteristics of device A based on active layer at 850 nm laser, **b** energy level diagram of device A

Fig. 4 **a** *I-V* characteristics of device B based on active layer at 850 nm laser, **b** Energy level diagram of Device B

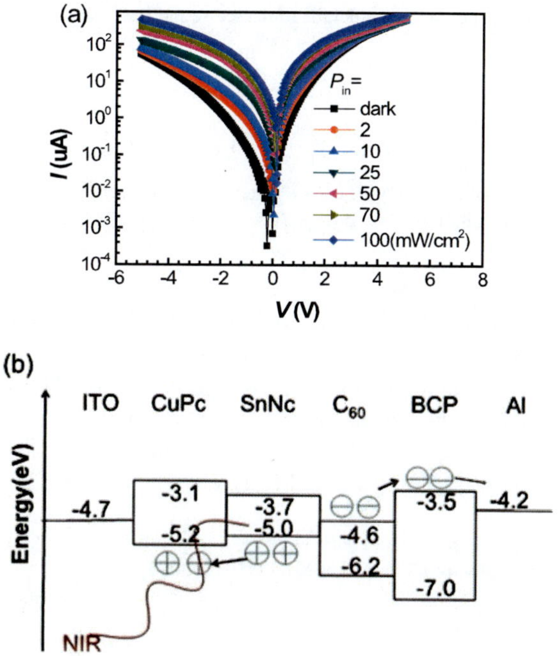

anode at the reverse bias voltage. As a result, Device A shows a high dark current and a low photosensitivity.

On the behalf of increasing the device performance, CuPc/SnNc/C_{60}-based OPD was fabricated by inserting CuPc electron blocking layer between the ITO anode and SnNc NIR sensitive layer (Fig. 1b). *I-V* characteristics of Device B based on active layer at 850 nm laser in the dark and under NIR illumination is described in Fig. 4. Device B (I_{dark}= 51.76 uA) has a lower dark current comparison with Device A (I_{dark}= 90.83 uA) under 100 mW/cm^2 light. The current reaches 469.53 uA at the reverse bias voltage of -5 V, which is ~9 times of the dark current. As shown in Fig. 4b, a high electron injection barrier of 1.6 eV exists at the interface of ITO anode and CuPc electron blocking layer, which effectively reduce the electron injection from the anode, resulting in effectively reduced the dark current and improved the device of photosensitivity. As summarized in Table 1, performance details of Device B about R, $P_{\max}$, EQE and $D*$ reach 401.22 mA/W, 10,978.91, 58.64% and 1.21×10^8, which was 3.3, 4.6, 3.3 and 4.2 times of Device A, respectively.

4 Conclusion

In conclusion, NIR OPD based on SnNc/C_{60} bi-layer and CuPc/SnNc/C_{60} tri-layer PHJs were fabricated and studied. The results showed that CuPc/SnNc/C_{60} tri-layer

Table 1 Performance details of device A and B

	I^{a}_{dark} (uA)	R^{a} (mA/W)	P^{b}_{max}	EQE^{a} (%)	$D*^{a}$ (Jones)
Device A	90.83	120.28	2384.14	17.58	2.89×10^{7}
Device B	51.76	401.22	10,978.91	58.64	1.21×10^{8}

[a]The results based on $V = -5$ V and P_{opt}=2 mW/cm^2
[b]The results based on device A at $V = -0.04$ V and device B at $V = -0.20$ V, and P_{opt}=100 mW/cm^2

PHJ photodiode superior performance compared with SnNc/C_{60} bi-layer PHJ photodiode. Under 850 nm light illumination, CuPc/SnNc/C_{60} tri-layer PHJ photodiode shows a large photoresponsivity of 401.22 mA/W, which is about 2.3 times more than SnNc/C_{60} bi-layer PHJ photodiode under the same conditions. The improved performance of device can be attributed to the enhanced electron injection barrier at the interface of anode and CuPc electron blocking layer.

Acknowledgements This work was supported by the Zhejiang Provincial Natural Science Foundation of China Grant No. LQ19F040003.

References

1. Källhammer JE (2006) Imaging: the road ahead for car night-vision. Nature Photonics, pp 12–13
2. Gao XH, Cui XY, Levenson RM, Chung LWK, Nie S (2004) In vivo cancer targeting and imaging with semiconductor quantum dots. Nat Biotechnol 22(8):969–976
3. Peng YQ, Lv LW, Yao B, Fan GY, Chen DQ, Gao PJ, Zhou MQ, Wang Y (2013) High performance near infrared photosensitive organic field-effect transistors realized by an organic hybrid planar-bulk heterojunction. Org Electron 14(4):1045–1051
4. Kim DY, Sarasqueta G, So F (2009) SnPc:C60 bulk heterojunction organic photovoltaic cells with MoO3 interlayer. Sol Energy Mater Sol Cells 93(8):1452–1456
5. Rand BP, Xue J, Yang F, Forrest SR (2005) Organic solar cells with sensitivity extending into the near infrared. Appl Phys Lett 87(23):233508
6. Kim DY, Song DW, Chopra N, Somer PD, So F (2010) Organic infrared upconversion device. Adv Mater 22(20):2260–2263
7. Liang YL, Lv WL, Luo X, He L, Xu K, Zhao F, Huang FB, Lu FP, Peng YQ (2018) A comprehensive investigation of organic active layer structures toward high performance near-infrared phototransistors. Synth Met 240:44–51
8. Kao PC, Chu SY, Huang HH, Tseng ZL, Chen YC (2009) Improved efficiency of organic photovoltaic cells using tris (8-hydroxy-quinoline) aluminum as a doping material. Thin Solid Films 517(17):5301–5304
9. Takanashi Y, Takahata K, Muramoto Y (1999) Characteristics of InAlAs/InGaAs highelectron-mobility transistors under illumination with modulated light. IEEE Trans Electron Devices 46(12):2271–2277
10. Rauch T, Bŏerl M, Tedde SF, Fürst J, Kovalenko MV, Hesser G, Lemmer U, Heiss W, Hayden O (2009) Near-infrared imaging with quantum-dot-sensitized organic photodiodes. Nat Photonics 3(6):332–336

11. Wang X, Li H, Su Z, Fang F, Zhang G, Wang J, Chu B, Fang H, Wei Z, Li B, Li W (2014) Efficient organic near-infrared photodetectors based on lead phthalocyanine/C_{60} heterojunction. Org Electron 15(10):2367–2371

The Manipulation of Two Types of Particles by a Radially Polarized Bessel-Gaussian Beam with Phase Modulation

Pengjie Sun, Jinsong Li, and Hongjie Ma

Abstract According to vector diffraction theory, the intensity of radially polarized Bessel-Gauss beams through the high numerical aperture(NA) objective with phase modulation system is numerically investigated. The intensity of the beam at the focal area is calculated. And the possibility of manipulating gold particles and bubbles is discussed by adjusting the phase distribution of the phase plate. By using the Rayleigh scattering theory, we calculated the gradient force, absorption force and scattering force of the light beam on the gold particles and bubbles with modulating phase, respectively. Under the thermal disturbance of the surrounding environment, the particles will do Brownian motion, and the generated Brownian force will have a certain influence on the capture stability, so we also calculated and analyzed the capture stability of the two different particles by the beam. The result of the simulation show that the radially polarized Bessel-Gauss beams is suitable for manipulating two kinds of particles by phase modulation.

Keywords Vector diffraction theory · Radially polarized Bessel-Gaussian beam · Phase modulation · Manipulate

1 Introduction

In 1970, researchers at bell LABS first discovered that particles could be captured and manipulated using the capture capability of highly focused beams [1]. Until 1986, the first real optical tweezers system was reported by Ashkin [2]. Since then, optical tweezers have been a key technology for the capture, characterization and manipulation of atoms and nanoscopic particles [3–6]. The force in optical tweezers is divided into two different contributions: one is the gradient force proportional to

P. Sun · J. Li (✉) · H. Ma
College of Optical and Electronic Technology, China Jiliang University, Hangzhou, People's Republic of China
e-mail: lijinsong@cjlu.edu.cn

Y. Peng and X. Dong (eds.), *Proceedings of 2019 International Conference on Optoelectronics and Measurement*, Lecture Notes in Electrical Engineering 726,
https://doi.org/10.1007/978-981-33-4110-4_8

the intensity of the light field, which is responsible for the capture of the particles; the other is the absorption force and scattering force, they have dissipative nature, not conducive to the particle capture. With a high NA objective focusing system, the radially polarized beam produces many unique advantages, such as smaller spot size and strong longitudinal field composition [7], which has become the focus of research. However, research on most radially polarized beams has focused on the Bessel-Gaussian beam mode or the Laguerre-Gaussian beam mode. Nie et al. designed the three-belt phase filter and realized the manipulation of two particles with radial polarized Laguerre-Bessel-Gaussian beam [8].

In this paper, we show that using a radially polarized Bessel-Gaussian beam to manipulate the two kinds of particles by adjusting the phase distribution of the phase plate according to vector diffraction theory. And we calculated the radiation force on the particle.

2 Theoretical Analysis

In Fig. 1, the system we investigated is depicted. We placed a circular phase plate in front of the objective lens, and this plate can change the phase of the incident beam. Then the modulated beam will focus by this lens.

In the high NA objective lens focusing system, the radially polarized Bessel-Gaussian beam can be described as:

$$l(\theta) = J_1\left(\frac{2\beta \sin\theta}{\sin\alpha}\right)\exp\left(-\frac{\beta^2\sin^2\theta}{\sin^2\alpha}\right) \quad (1)$$

In the above equation, β is the ratio of the radius of pupil and beam waist, $\alpha = \arcsin(NA/n)$ is the maximum of the convergence angle θ, and J_1 represents the first order Bessel function, respectively.

Based on Richards and Wolf methods [9, 10], at the focal point, the electric field intensity $E(r, z)$ of the radially polarized Bessel-Gauss beam focused by the high NA lens with an additional phase plate can be expressed as

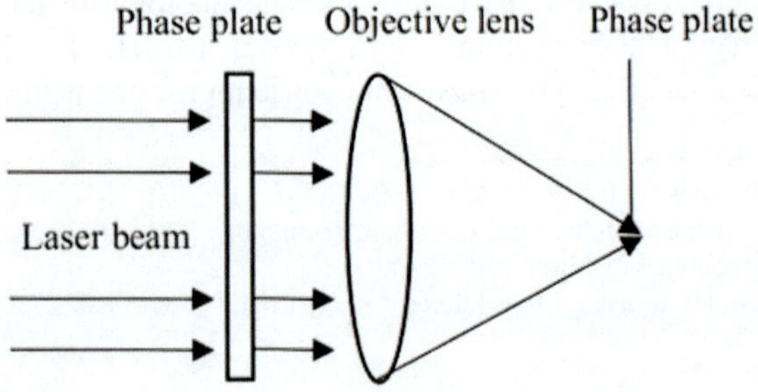

Fig. 1 Phase-modulated focusing optical system

$$E_r(r, z) = A \int_0^{\alpha} \sqrt{\cos\theta} \sin(2\theta) l(\theta) T(\theta) J_1(kr \sin\theta) \exp(ikz \cos\theta) d\theta \quad (2)$$

$$E_z(r, z) = 2iA \int_0^{\alpha} \sqrt{\cos\theta} \sin^2\theta l(\theta) T(\theta) J_0(kr \sin\theta) \exp(ikz \cos\theta) d\theta \quad (3)$$

According to Maxwell's equations, the relationship between magnetic and electric fields is as follows

$$H = -\frac{i}{k}\sqrt{\frac{\varepsilon}{\mu}} \nabla \times E \quad (4)$$

By substituting Formulas (2) and (3) into Formula (4), the magnetic field intensity of radially polarized Bessel-Gauss beams focused by high NA lenses should be written as follows

$$H_{\varphi}(r, z) = \frac{2An}{\mu_0 c} \int_0^{\alpha} \sqrt{\cos\theta} \sin(\theta) l(\theta) T(\theta) J_1(kr \sin\theta) \exp(ikz \cos\theta) d\theta \quad (5)$$

where r and z represent the direction in the cylindrical coordinates, $A = Efn^{1/2}/\lambda$, E is the amplitude determined by the incident beam energy, λ and f are incident wavelength and focal length, respectively. $k$denotes the wave number, which relationship with λ is $k = 2\pi/\lambda$, c is the light speed, μ_0 is the permeability of vacuum, $T(\theta)$ represents the phase transmission function.

In the focal region, the intensity distribution and average Poynting vector of the beam are expressed as

$$I(r, z) = |E_r(r, z)|^2 + |E_z(r, z)|^2 \quad (6)$$

$$\langle S \rangle = \frac{1}{2}\{\mathrm{Re}[E_r(r, z) H_{\varphi}^*(r, z)]\vec{e_z} - \mathrm{Re}[E_z(r, z) H_{\varphi}^*(r, z)]\vec{e_r}\} \quad (7)$$

In Formula (9), $\vec{e_r}$ is the unit vectors in r direction, and $\vec{e_z}$ is the unit vectors in z direction.

In an optical field, the Rayleigh particle can be considered a point dipole, because of its radius is smaller than the wavelength of the incident beam. And the point dipole polarizability α is expressed as

$$\alpha = 4\pi a^3 \varepsilon_1(\varepsilon_2 - \varepsilon_1)/(\varepsilon_2 + \varepsilon_1) \quad (8)$$

where a represents the radius of the Rayleigh particle, ε_1 is the dielectric constants of the particle, and ε_2 is the dielectric constants of the medium surrounding the particle. Based on Rayleigh scattering theory [11], the gradient, absorption and scattering forces can be written as

$$F_{grad} = \frac{1}{4}\text{Re}(\alpha)\varepsilon_0 \nabla I(r, z) \tag{9}$$

$$F_{\text{abs}} = n\langle S\rangle C_{abs}/c \tag{10}$$

$$F_{scat} = n\langle S\rangle C_{scat}/c \tag{11}$$

where ε_0 is the permittivity in a vacuum, $C_{scat} = k^4|\alpha|^2/6\pi$ is the scattering cross sections, $C_{abs} = kn\ \text{Im}(\alpha)/\varepsilon_1$ is the scattering cross sections. Re and Im represent real and imaginary parts, respectively.

In trapping force by the beams, the particles are subjected to three forces: gradient force, absorption force and scattering force. Using Formulas (6)–(11), the radially Bessel-Gaussian beam radiation force on the particle can be calculated.

3 Results and Discussion

In the simulation calculation, the laser power is 500 mw, the wavelength λ is 1047 nm, and the manipulated particle is placed in water(n = 1.332).$NA = 0.95$, $\beta = 1.9487$.

To realize the manipulation of bubbles, we add a phase plate to the system. The transmission function $T(\theta)$ of it is written as

$$T(\theta) = \begin{cases} \exp(-i\pi/2), & \text{for } 0 \leqslant \theta \leqslant \theta_1 \\ \exp(i\pi/2), & \text{for } 0 \leqslant \theta \leqslant \alpha \end{cases} \tag{12}$$

θ_1 and α correspond to two radial positions r_1 and r_2, respectively. The bubble has a radius of 30 nm. And it has a refractive index is 1, less than 1.33 of water.

Here we give an example of adjusting the phase distribution, and choose two angles represented as $\theta_1 = 35.90°$ and $\alpha = 71.8°$, respectively. And we calculated the gradient force F_{grad} and sum of the scattering force and absorption force $F_{scat} + F_{abs}$ acted on the bubble by the radially polarized Bessel-Gaussian beam as shown in Fig. 2.

Comparing Fig. 2(a) with Fig. 2c, it is found that the gradient force acting on bubbles in the r direction is much bigger than the sum of absorption force and scattering force. Comparing Fig. 2b with Fig. 2d, we can get the same situation in z direction.

Now, to manipulate the gold particle, which has a higher refractive index than water, we add another phase plate to the system. The transmission function $T(\theta)$ of it is expressed as

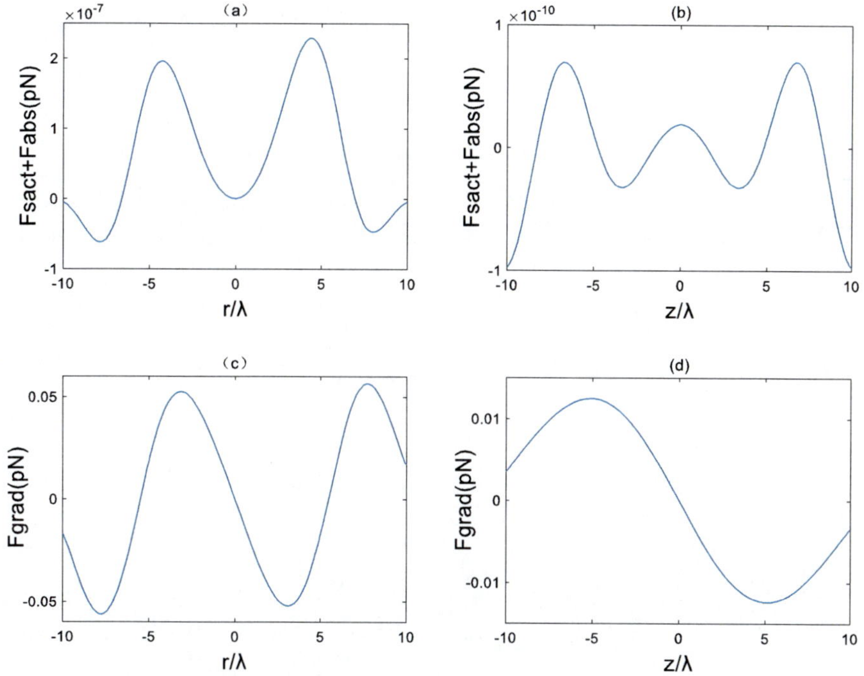

Fig. 2 The radially polarized Bessel-Gaussian beam acts on the bubbles. The sum of scattering force and absorption force, the gradient force along the **a**, **c** r direction and **b**, **d** z direction

$$T(\theta) = \begin{cases} 1, & \text{for } 0 \leqslant \theta \leqslant \theta_1 \\ \exp(i\pi/3), & \text{for } 0 \leqslant \theta \leqslant \alpha \end{cases} \tag{13}$$

The gold particle has a radius of a = 19.1 nm, and its permittivity is $\varepsilon_2 = -54+5.9i$. The corresponding dielectric constant of water is 1.7689. It is obvious that the gold particles refractive index is greater than that of water.

We here give an example of adjusting the phase distribution that $\theta_1 = 62.83°$ and $\alpha = 71.8°$. In Fig. 3, the force on the gold particle we calculated is shown. We can see from Fig. 3 that the gradient force on the gold particles is still far greater than the sum of the absorption force and scattering force in all directions. When the force is positive, the particle is subjected to a forward thrust in the positive direction, however, when the force is negative, the particle is subjected to a pulling force in the negative direction. And from Fig. 2a, c, we could get that the gradient force F_{grad} and the sum of the scattering force and the absorption force $F_{scat} + F_{abs}$ of the bubble in the negative direction of the r-axis are positive, causing the bubble to move to the focus along the positive direction in r-axis. The bubbles at the focus are not affected by force. In the positive half-axis of r, F_{grad} is negative, causing the bubble to move in the negative r direction. At the same time, $F_{scat} + F_{abs}$ is positive, pushing the particle to move in the positive direction of the r-axis. Because of F_{grad} is greater

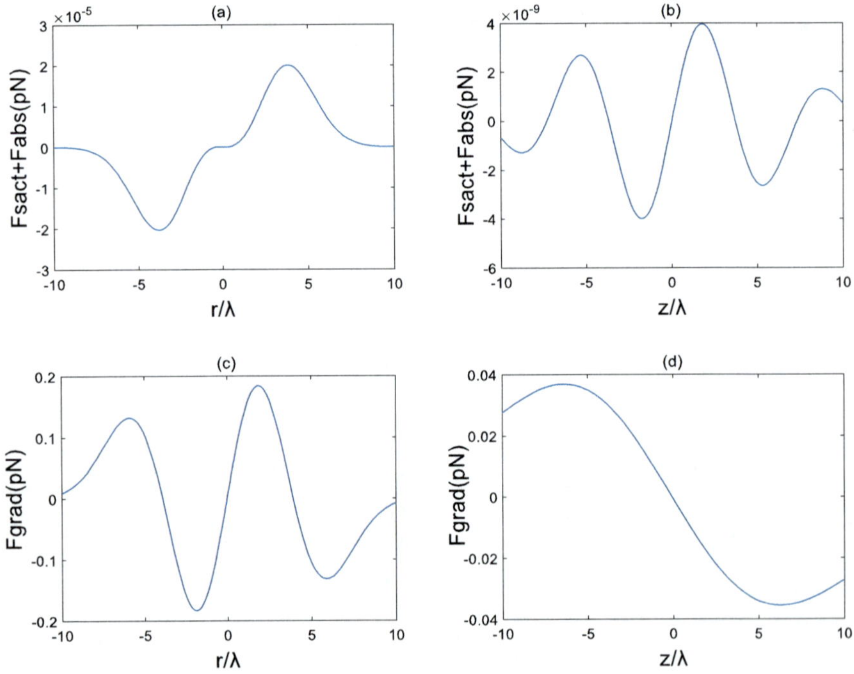

Fig. 3 The radially polarized Bessel-Gaussian beam acts on the gold particle. The sum of scattering force and absorption force, the gradient force along the **a**, **c** r direction and **b**, **d** z direction

than $F_{scat} + F_{abs}$, the bubble will reach the force balance at the focus. Since F_{grad} is much larger than $F_{scat} + F_{abs}$ in the z direction, the same force condition applies to the z direction as can be seen from Fig. 2b, d. The force of the gold particles is shown in Fig. 3. We can see from Fig. 3a, c that in the negative half-axis of the r-axis, F_{grad} and $F_{scat} + F_{abs}$ are both negative, while in the positive half-axis, all three forces are positive, this means that the gold particle will be subjected to a thrust that causes it to move away from the focus and move in the negative and positive directions, respectively. But at the focus, the force is zero and the gold particles are unstressed. In the z direction, the gold particles will stabilize at the focus under the influence of three forces, as shown in Fig. 3b, d.

To achieve stably manipulate, some stability criteria need to be satisfied. Firstly, F_{grad} should be large enough than $F_{scat} + F_{abs}$, namely, $R = F_{grad}/(F_{abs} + F_{scat}) > 1$, and R is the stability criterion. In order to more conservative estimates, we use $R = \left(F_{grad}\right)_{\max}/\left[(F_{abs} + F_{scat})\max\right] > 1$ [12] to estimate stability. After simple calculations. We have obtained R for bubbles in the r and z directions of 78,986 and 7.88×10^7, respectively, and for gold particles, R in the z direction is 9.67×10^6. The results show that the radial-polarized Bessel-Gaussian beam can capture bubbles and laterally move the gold particles by adjusting the lens of the additional phase plate.

Secondly, the particles will generate kinetic energy due to Brownian motion, so in order to be able to manipulate the particles stably, the potential well produced by the gradient force should be deep enough to overcome this energy. A commonly accepted criterion for particle trapping stability is $R_{thernmal} = \exp(-U_m/k_B T) \leqslant 1$ [2], where U_m is the maximum potential depth given by $|\mathrm{Re}(\alpha)\varepsilon_0 I(r, z)_{\max}/2|$, k_B is the Boltzmann constant, and T represents the temperature of the medium around the particle. We assume the temperature is 300 K, for the air bubble, $R_{thernmal}$ is calculated to be 3.17×10^{-2}, for the gold particle $R_{thernmal}$ is 3×10^{-3}. The results show that the radially polarized Bessel-Gaussian beam can overcome Brownian motion to manipulate two different types of particles.

4 Conclusion

In summary, based on vector diffraction theory, the radiation force of a radially Bessel-Gauss beam on bubbles and gold particles is calculated by changing the phase distribution of the additional phase plate lens. The gradient force generated by the beam on the bubble is far greater than the sum of scattering and absorption forces and the potential well depth generated by the gradient force can overcome the kinetic energy of the bubble. For gold particles, the beam is capable of stably controlling the gold particles in the direction of propagation and move horizontally in the focal plane, respectively. Therefore, by using the phase-modulated system, a radially polarized Bessel-Gaussian beam can steadily manipulate two different particles.

References

1. Ashkin AF (1970) Acceleration and trapping of particles by radiation pressure. Phys Rev Lett 24(4):156
2. Ashkin A, Dziedzic JM, Bjorkholm JE (1986) Observation of a single-beam gradient force optical trap for dielectric particles. Opt Lett 11(5):288–290
3. Foot CJ (2015) Atomic physics. Oxford University Press, Oxford
4. Fan C, Liu Y, Wang X (2018) Trapping two types of particles by using a tightly focused radially polarized power-exponent-phase vortex beam. J Opt Soc Am A 35(6):903–907
5. Zhang H, Li J, Cheng K (2017) Trapping two types of particles using a focused partially coherent circular edge dislocations beam. Opt Laser Technol 97:191–197
6. Maragò OM, Jones PH, Gucciardi PG (2013) Optical trapping and manipulation of nanostructures. Nat Nanotechnol 8(11):807
7. Dorn R, Quabis S, Leuchs G (2003) Sharper focus for a radially polarized light beam. Phys Rev Lett 91(23):233901
8. Nie Z, Shi G, Li D (2015) Tight focusing of a radially polarized Laguerre–Bessel–Gaussian beam and its application to manipulation of two types of particles. Phys Lett A 379(9):857–863
9. Richards B, Wolf E (1959) Electromagnetic diffraction in optical systems. II. Structure of the image field in an aplanatic system. Proc R Soc Lond Ser A Math Phys Sci 253:358–379
10. Focusing of high numerical aperture cylindricalvector beams (2000) K.S. Youngworth, F., T.G. Brown, S. Opt Express 7:77–87

11. Zhang Y, Ding B, Suyama T (2010) Trapping two types of particles using a double-ring-shaped radially polarized beam. Phys Rev A 81(2):023831
12. Zhan QF (2004) Trapping metallic Rayleigh particles with radial polarization. Opt Express 12:3377–3382

Application of Light Wave in Surface Science and Surface Treatment Technology

Gaohui Zhang, Kai Wang, and Yang Wang

Abstract The history of optical development reflects the history of science and technology to some extent, the origin of light and matter is a very fundamental issue in physics, corpuscular and Wave theory of light has played a very important role in promoting the development of modern science and technology. This paper is based on material surface science and technology, discusses the precise measurement and evaluation of surface science and technology by means of light refraction and interference. Set forth the application of high energy, high monochromaticity and high directivity laser in surface science and surface treatment technology of materials such as laser cladding and laser surface alloying, points out that light wave is an advantageous weapon for surface science and surface treatment technology progress, it provides conditions for improving material surface properties, expanding material application field and prolonging material service life. It is also predicted that the optical progress in the future will continue to be a favorable support for the development of material surface science and surface treatment technology.

Keywords Light · Laser · Surface science · Surface treatment

1 First Section

Light is composed of photons and has wave-particle duality, its essence is electromagnetic wave. Light waves have physical properties such as interference and diffraction due to the wave character of light. Its several thousand angstroms of wavelength provide the conditions for accurate measurement. Since July 1960, Dr. T. Mayman of Hughes Aircraft Company in New York announced that he developed the world's first laser on May 15th, generating the first laser [1].Laser is characterized by its high

G. Zhang (✉) · K. Wang · Y. Wang
Research Institute of Surface Physics, China Jiliang University, Hangzhou, People's Republic of China
e-mail: ghzhanggh@163.com

Y. Peng and X. Dong (eds.), *Proceedings of 2019 International Conference on Optoelectronics and Measurement*, Lecture Notes in Electrical Engineering 726,
https://doi.org/10.1007/978-981-33-4110-4_9

energy, high monochromatism, high brightness, high sensitivity and high accuracy, and the wave band is continuously adjustable, including infrared, visible light, ultraviolet to X-ray bands, revolutionized the science of optics, and lasers changed the available quantities of physics on a large scale. The birth of the laser has added a new way for human beings to understand and transform the world, it has had and will continue to have a significant and profound impact on the contemporary information society. In measuring distance, laser drilling, cutting, scratching, welding, integrated circuit packaging, resistance fine-tuning, chip cleaning and so on have been inseparable from laser. Laser therapy technology involves almost every subject, and plays an irreplaceable role in urology, cardiovascular, orthopedics, dentistry and other specialties. Up to now, laser technology has been widely used in national defense, industry, chemistry, medical and other fields [1–5]. In all fields of scientific research, laser has become a powerful scientific research means. Especially the advent of femtosecond laser pulses has made microfabrication even further. Laser technology also plays an important role in surface science and surface treatment technology.

2 Optical Measurement

2.1 *Measurement of Surface Profile*

Surface three-dimensional micro-topography has the most direct impact on the evaluation of many technical properties of engineering parts, and the surface three-dimensional evaluation parameters are paid more and more attention because they can reflect the surface characteristics of parts more comprehensively and more truly and measure the surface quality, so the measurement of three-dimensional surface microtopography becomes more and more important. Through the measurement of three-dimensional topography, the surface quality can be evaluated comprehensively, and then the rationality of the processing method and design requirements can be confirmed. In this way, high-quality surface can be processed by guiding the processing and optimizing the processing technology to ensure the realization of parts' use functions. Optical 3D surface profilometer is a detection instrument for sub-nanometer measurement of the surface of precision devices and materials. It is based on the principle of white light interference technology, combined with precision z-scan module and 3D modeling algorithm to perform non-contact scanning on the surface of the device and establish 3D image of the surface, and using the system software to process and analyze the 3D image of the device surface, obtain the parameters reflecting the surface quality of the device, the optical detection instrument for 3D measurement of the device surface topography is realized. It is used in semiconductor manufacturing, precision parts, optical processing, micro-nano materials and other ultra-precision processing also in industries, aerospace, national defense, scientific research institutes and other fields [6]. It can be used to measure the roughness, flatness, microscopic geometric contour and curvature of all kinds of objects from super smooth to rough, low reflectivity to high reflectivity, and from nanometer to micron.

2.2 Measurement of Surface Profile

The surface roughness is the roughness of the surface with small spacing and small peak valley. The surface roughness is closely related to the matching properties, wear resistance, fatigue strength, contact stiffness, vibration and noise of mechanical parts, which are generally formed by the processing methods adopted and other factors. It has important influence on the service life and reliability of mechanical products. Measurement of roughness is of great significance in understanding the surface characteristics and service life of products. The traditional measurement method is stylus method, which has certain requirements on the surface roughness, and the accuracy is not too high. The study of surface roughness of metal films by optical methods such as ellipsometry, interferometry, stylus method and scattering has been paid more and more attention because of its timeliness and non-destructiveness: ellipsometry is the method of measuring the change of the reflected light polarization of the film deposited on the substrate by ellipsometry to obtain the information about the thickness of the film and the optical parameters quantitatively. Theoretically, the measurement sensitivity of the ellipsometry technique can reach the order of 0.1 nm. Deng Liwen and Wang Gongming use surface plasmon spectroscopy (SPS) to determine the surface roughness of metals [7, 8],The roughness measurement by comparison method is based on the comparison between the visual and tactile sense and the surface under test, judging that the roughness of the surface under test is equivalent to that value, or measuring the change of reflected light intensity to evaluate the surface roughness; Light-section method of roughness is that the light band formed by the light passing through the slit is projected onto the measured surface, and the surface roughness is measured by the contour curve formed by the line of intersection with the measured surface; Interferometric measurement of roughness uses the principle of light wave interference to display the shape error of the measured surface as an interference fringe pattern, and uses a microscope with a high magnification (up to 500 times) to magnify the microscopic part of these interference fringe to measure, to get the measured surface roughness.

2.3 Microtopography Test

The microstructure of material surface is a very important index to measure the state of material surface. The microstructure of the material surface can be investigated to assess the basic composition of the material surface, which has certain significance for the roughness and porosity of the material investigation, so as to provide a basis for studying the wear resistance and corrosion resistance of materials, etc. Optical

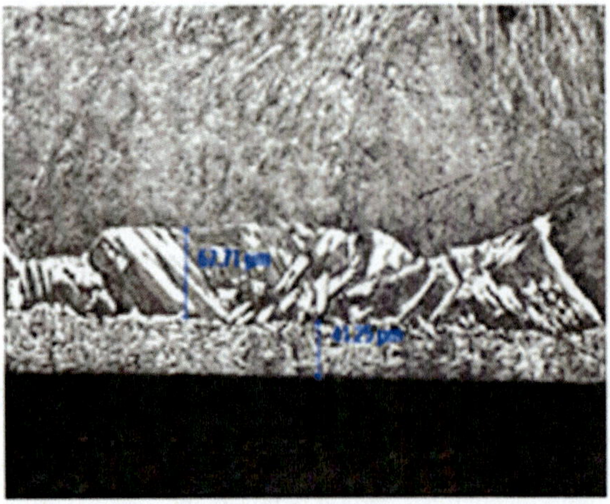

Fig. 1 Hydrogen-free carburization of Ti alloy

microscopy is the most direct means to obtain the surface morphology of materials, and the magnification can reach thousands of times. Figure 1 is the microscopic morphology of the titanium alloy surface after hydrogen-free carburization obtained by the optical microscope [9, 10], the thickness and existing state of the carburizing layer after the carburizing of titanium alloy can be clearly understood through the photos.

2.4 Laser Scattering Measurement of Liquid Surface Tension

There are thermally excited capillary waves on the liquid surface. The wavelengths (micron order) and amplitude (nano order) of these waves are very small. For the light incident on the liquid surface, these waves are like moving gratings, which will cause incident light to occur Brillouin scattering. The properties of surface waves are related to the surface tension of liquid, the surface tension of liquid can be obtained by analyzing the scattered light [11]. Compared with the traditional method, the method of measuring liquid surface tension by laser scattering is non-contact, fast and can be used for limit measurement.

2.5 Measurement of Optical Temperature

Temperature is almost the most important parameter in the heat treatment of materials, the high or low temperature and the speed of temperature change have a great influence on the properties of materials. However, it has become a difficult problem in material science to obtain accurate temperature under the restriction of conditions.

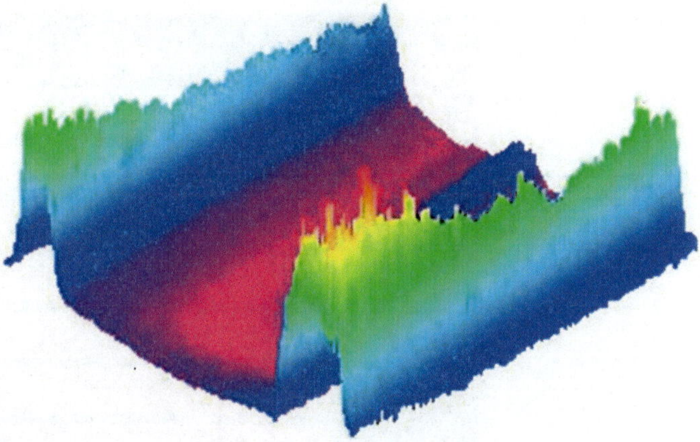

Fig. 2 The temperature distribution of the plasma in the double glow discharge

All objects whose temperature is above absolute zero constantly emit infrared radiation energy to the surrounding space. The amount of infrared radiation energy of an object and its distribution by wavelength are very closely related to its surface temperature. By measuring the infrared energy radiated by an object, its surface temperature can be accurately measured. Infrared temperature measurement is the infrared radiation energy emitted by the object into electrical signals, infrared radiation energy and the size of the object itself corresponding to the temperature, according to the size of the conversion into electrical signals, you can determine the temperature of the object. With the development of digital technology, it has become a reality to use digital camera to obtain the thermal effect picture of temperature and then obtain the temperature field according to digital imaging. Figure 2 shows the temperature field in the double glow ion metallization technique calculated by professor zhang pingze using digital imaging of light [12]. Figure 3 is the color temperature diagram.

2.6 Flame Resistance Test

The femtosecond laser, which works in the form of pulse, has the characteristics of extremely short pulse duration, extremely high pulse instantaneous power and extremely strong focusing electromagnetic field. The beam characteristic of femtosecond laser pulse determines that it is the most ideal tool in laser micromachining. The main feature of femtosecond laser processing is that laser pulses can inject energy into the processed area in less time than the thermal diffusion of matter. That is to say, the laser pulse has ended before the thermal diffusion of the energy irradiated into the material, so there is no thermal diffusion problem in the irradiated area, so that extremely high thermal efficiency finishing can be obtained.

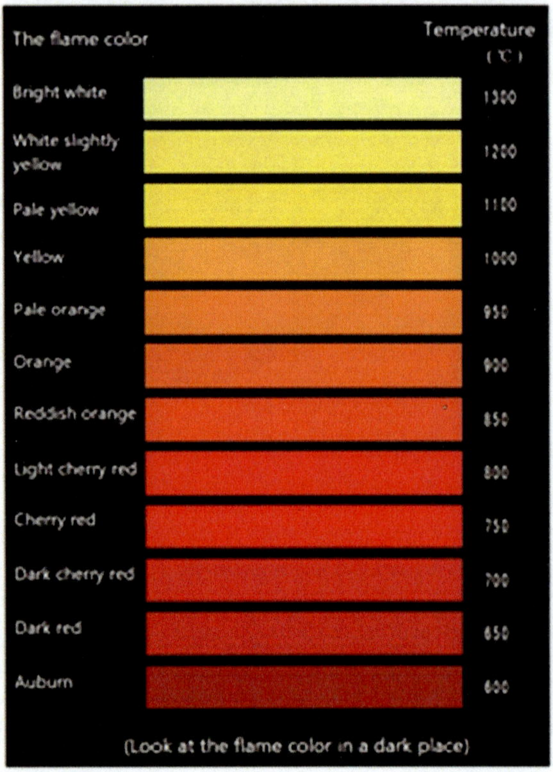

Fig. 3 Color temperature diagram

It is a difficult problem to test the flame retardancy of materials by simulating the burning scenario. The femtosecond laser is used to perform the cauterization experiment on the surface of the material. It is convenient and accurate to judge the ablation performance based on the traces of the cauterization. Figure 4 shows a laser ablation experiment using a mode-locked titanium sapphire laser Mai Tai HP after infiltration copper plating on a titanium alloy Ti6Al4V. The experimental conditions first use anhydrous methanol to clean the sample surface so that the sample surface has the same absorbance. Experimental conditions, then at room temperature and pressure, laser power 2.60 w, monopulse pulse width 100 fs, central wavelength 780 nm, repetition rate 80 MHz, laser ablation 60 s. After the experiment, Olympus metallographic microscope was used to photograph the surface combustion morphology [13]. Compared with the samples without copper plating, the center ablation area was significantly smaller, indicating that the flame retardancy was improved.

Fig. 4 Ablation morphology

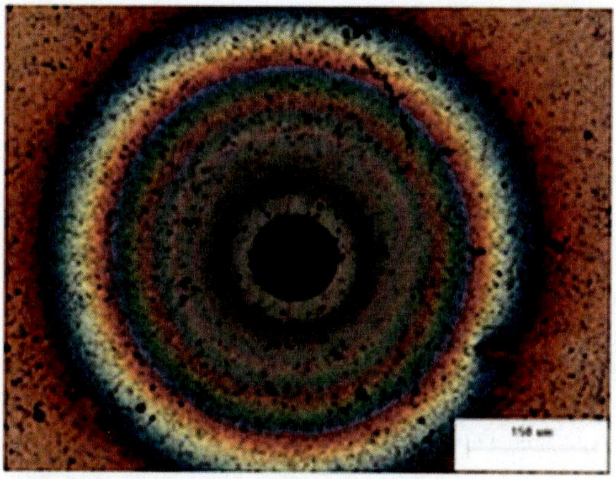

3 Laser Surface Treatment

Laser surface treatment is to use high power density of laser beam, using a laser of high energy, high directivity characteristics of the processing surface heat exchange quickly, heating surface in the form of non-contact, conduction cooled by means of material surface itself, in the processing of material surface to form certain thickness layer, improve material surface organization and structure, to improve the surface strength, wear resistance, corrosion resistance, fatigue resistance and a series of performance. The energy transfer is convenient, can have the selective local strengthening and the energy action concentration. Laser surface treatment has the advantages of short processing time, small impact area, small deformation, more significant effect, high speed, high efficiency.

3.1 Laser Quenching

The laser surface quenching technology began in the mid-1970s, using a laser beam of high-energy density to scan metal surfaces in non-contact ways, after the absorption of light energy, the temperature rises to above the austenitic phase transition temperature and below the melting point temperature. Such a fast temperature rise process is beneficial to the formation of austenite nuclei, which can obtain fine austenite grains, and then perform rapid heat transfer and rapid cooling by itself. Because the heating time is short and the cooling rate is too fast, the austenite is too late to homogenize, resulting in uneven distribution of carbon and alloy elements, the difference between the fine grained austenite and the austenite structure composition is conducive to finally obtaining fine and high hardness martensite. Make the surface of the workpiece extremely compressive stress, greatly improve the fatigue strength of the material [13].The surface hardness of the workpiece after laser hardening treatment is high, 15–20% higher than that of conventional quenching, and the surface condition is good, which can be used as the final processing procedure.

3.2 Laser Etching

Laser etching technology is a flexible processing technology that combines pulsed laser etching processing with advanced CNC system. It uses short pulses and high peak power laser pulses to rapidly and locally heat up, melt and even vaporize the surface of the material to be etched, which can achieve accurate removal of surface materials. Due to the pulse action time is very short (ns scale), the thermal influence zone formed in the etched material is very shallow (micron scale), laser etch can control the etch depth through energy density, it also has characteristics such as non-contact and non-introduction stress. Secondly, the multi-degree of freedom laser etching machining head controlled by computer program can achieve accurate positioning and graphic etching on any complex surface, which has high machining and positioning accuracy and can complete the etching of curve graphics at one time. Lanzhou institute of space technology and physics began to adopt the coating/laser etching technology to carry out the research work of surface metal thin film graphic structure etching, Up to now, the physical process model of laser etching process for material combination systems with different thermal properties has been established, it also solves technical problems such as low-temperature deposition technology of metal thin films on composite materials, large and precise cross-scale laser etching equipment, and related processes [14].The further development direction of laser etching technology is to improve the ability of etching thick metal thin films and expand the application in micromachining. Laser etching is a method of direct etching. It is also possible to further improve the accuracy and resolution of laser etching, so that it may be replace photolithography to a certain extent in the field of micromachining.

3.3 Laser Cladding

The cladding alloy is first deposited on the surface of the material by bonding, spraying, electroplating, etc. Then the alloy is melted by laser beam, and then the material itself is rapidly cooled and solidified. The laser surface melting technology is used to achieve the rapid heating and cooling of the local surface of the material so as to obtain a very fine non-equilibrium rapid solidification structure. The formation of optical cladding layer is a complex physical and chemical process. It can effectively reduce cracks, holes and inclusions in the cladding layer, promote grain refinement and improve the microstructure uniformity of the cladding layer. Laser cladding metal-based ceramic composites are composed of oxides, carbides, nitrides, borides, silicides and other high melting point rigid ceramic materials compounded with metals. Laser cladding metal-based ceramic composites are made of various high melting point hard ceramic materials such as oxides, carbides, nitrides, borides and silicides and are compounded with metals, It can combine the high strength, toughness, good technological performance of metal materials with the excellent wear resistance, corrosion resistance, high temperature resistance and chemical stability of ceramic phase, and often used to prepare high temperature oxidation resistant coating, wear-resistant coating, bioceramic coating, etc.

3.4 Laser Alloying

A high energy laser beam is used to mix and melt the alloy coating with the base metal surface. In a very short time, the alloy layer with different chemical composition and structure surface is formed, so that the surface of the matrix has specific alloy composition. Because of the fast heating speed of laser, the melting zone and the heat-affected zone are very small. The alloying elements are completely dissolved in the surface layer, and the composition of the modified layer is very uniform, not sensitive to chapping and peeling. Alloying of inaccessible and localized areas; With deep focusing, uniform alloying depth can be obtained on irregular parts; The power density and heating depth can be controlled accurately to reduce deformation; Save a lot of valuable elements. In addition, laser alloying can obtain a surface strengthening layer with fine grains, high hardness and firm combination with the matrix [18, 19].For example, after laser alloying and strengthening, an excellent wear resistant layer is formed on the surface of 718 steel, which greatly improves the service life of the mold.

3.5 Laser Surface Amorphous

Using laser beam to scan the metal surface continuously and quickly to melt the surface metal, and cooled below the crystallization temperature at a rate greater than the critical cooling rate to form amorphous on the metal surface, also known as metallic glass [20, 21]. Laser amorphous also known as laser glazing. The microhardness of the laser amorphous layer is much higher than that of the corresponding components. In addition, amorphous treatment can reduce surface component segregation, eliminate surface defects, with good toughness, wear resistance, corrosion resistance.

Also, laser cleaning, laser shock intensification, laser vapor deposition, laser repair and other technologies have good applications in surface science and technology, which will not be described here due to limited space.

4 Conclusion

Science and technology are advancing with each passing day. Advanced technology and new fields are emerging. Extreme working conditions put higher and higher demands on work components. New materials keep popping up. As a result, the application of light wave in the field of surface science and surface treatment technology has raised new requirements and provided a new research direction and development space. Although laser processing technology has the disadvantages of expensive equipment and processing complex parts that require big data automatic control support, but light wave technology will continue to inherit its advantages with its unique performance characteristics, which will provide a good support and promotion for the development of surface science and surface treatment technology.

References

1. Spence DE, Kean PN, Sibbett W (1991) 60-fsec pulse generation from a self-mode-locked Ti: sapphire laser. Opt Lett 16(1):42–44
2. Wu YS, Li JM, Lu CH (2014) Influence of laser parameters on quality of laser direct marked data matrix symbols on aluminum alloy. J Mech Electr Eng 31(7):865–869
3. Zheng Y, Wang CL, Zhang GY (2014) Microstructure and corrosion resistance of laser cladding rare earth CeO2+Ni60 alloy coatings on 6061 Al alloy. Trans Mater Heat Treat 148–152
4. Xu P, Lin CX, Zhou CY et al (2013) Preparation and microstructure of 304 stainless steel layer by laser cladding. Trans Mater Heat Treatment 34(7):142–146
5. Zhang JH, Sun RL (2014) Research progress of laser cladding on surface of Titanium Alloy. Mater Rev 89–93
6. Nuttens T, Wulf AD, Bral L et al (2010) High resolution terrestrial laser scanning for tunnel deformation measurements. FIG Congress
7. Deng LW, Wang GM (1998) Surface plasmon spectroscopy for measuring the surface roughness of metallic films. Acta Optica Sinica 4(18):474–480
8. Sun DL, Wu Q, Liu CS et al (2008) Application of laser scanning confocal microscopy in roughness characterization of worn surface. Chin J Lasers 9(35):1409–1414
9. Zhang GH, Pan JD, He ZY et al (2005) Double glow plasma hydrogen-free carburizing on commercial purity titanium. J Wuhan Univ Technol Mater Sci 4(20):80–82
10. Zhang GH, Zhang PZ, Pan JD et al (2004) Double glow plasma carburizing without hydrogen of titanium alloy. Nonferrous Metal 2(1):4–6
11. Wang FK, Wu JT, Liu ZG (2006) Surface tension measurement system with surface laser light scattering method. J xi'an Jiaotong Univ 9(40):1006–1009
12. Zhang PZ, He ZY, Zhang GH et al (2008) Study on temperature distribution in double glow discharge zone. J Mater Sci Forum 575–578:1281–1286
13. Huang GQ, Zhang GH, Qiao XW et al (2010) Ablation character of the titanium alloy surface by magnetron sputtering copper. J China Univ Metrol 9(3):274–282
14. Yang LQ, Ding YX (2006) Research and development of laser hardening technology. Hot Working Process 35(4):68–70
15. Yang JP, Chen XK, Wu G (2011) Laser ablation and its pplication in fabrication of spacecraft antennas. Aerospace Manuf Technol 6(3):51–54
16. Ye XY, Shin YC (2014) Synthesis and characterization of Fe-based amorphous composite by laser direct deposition. Surf Coat Technol 239:34–40

17. Zheng Y, Wang CL, Gao Y et al (2014) Microstructure, friction and wear behaviors of laser cladding rare earth CeO2+Ni60 alloys coating on Al alloy surface. Chin J Rare Metals 5(38):800–806
18. Zhang W, Shi SQ, Chen YX (2011) The research of microstructure and property of laser alloying on the surface of 718 steel compared with nitriding. Appl Laser 117–119
19. Tian YS, Chen CZ, Chen LX, Huo QH (2006) Microstructures and wear properties of composite coatings produced by laser alloying of Ti6A14V with graphite and silicon mixed powders. Mater Lett 60(1):109–113
20. Inoue A, Takeuchi A (2011) Recent development and application products of bulk glassy alloys. Acta Mater 59(6):2243–2267
21. Basu A, Samant AN, Harimkar SP et al (2008) Laser surface coating of Fe-Cr-Mo-Y-B-C bulk metallic glass composition on AISI 4140 steel. Surf Coat Technol 202(12):2623–2631
22. Zhu GD, Wang SR, Cheng W (2019) Advances in the application of laser cleaning to metal surface treatment. Shandong Sci 32(4):38–45
23. Abboud JH (2013) Effect of processing parameters on titanium nitrided surface layers produced by laser gas nitriding. Surf Coat Technol 214:19–29
24. Ding YX, Yang LQ (2006) Research status and development of laser heat treatment. Mech Eng 1:19–21
25. Romanos GE, Gutknecht N, Dieter S et al (2009) Laser wavelengthsand oral implantology. Lasers Med Sci 24(6):961–970
26. Xie ZY, Pan YX (2003) The phase change mechanism and application of laser heat treatment. Mach Manuf Autom 4:38–42

Broadband Reversed Fano Switch Based on a Ring-Bus-Ring-Bus System

Qiqi Yuan, Huihui Zhang, Qiang Liu, and Qingzhong Huang

Abstract We have proposed and experimentally demonstrated a broadband thermo-optic switch in silicon-on-insulator. The switch is based on a ring-bus-ring-bus system, which can generate reversed Fano lineshapes in the transmission spectra. Then, the fabricated device exhibits a 3-dB-bandwidth more than 280 GHz, a maximum extinction ratio of 28 dB, and a slope rate of 60.8 dB/nm. In addition, the simulation results reveal that the switch is capable of switching data rates as high as 320 Gb/s with negligible signal distortion.

Keywords Integrated optics devices · Micro-optical devices · Coupled resonators · Optical switching devices

1 Introduction

Fano resonance with a sharp asymmetric lineshape has been observed and widely investigated in the atomic systems previously [1, 2]. Recently, analogue to Fano effect is also found in the integrated photonic devices. Owing to the sharp bandedge, Fano resonance has attracted substantial interests in the fields of low-power optical switching and modulating [2–4], high-sensitivity sensing [5], optical filtering and lasers [6]. Various structures have been demonstrated to implement a Fano resonance, such as waveguide-coupled-cavities [7], photonic crystals [2, 8], and metasurface. However, the optical Fano switch is still limited by the narrow bandwidth, which will cause significant spectral distortion and sideband attenuation for a relatively large data bandwidth. In other words, it will reduce the robustness and capacity of switches [9].

Q. Yuan · H. Zhang · Q. Liu · Q. Huang (✉)
Wuhan National Laboratory for Optoelectronics, Huazhong University of Science and Technology, Wuhan, People's Republic of China
e-mail: huangqz@mail.hust.eedu.cn

Y. Peng and X. Dong (eds.), *Proceedings of 2019 International Conference on Optoelectronics and Measurement*, Lecture Notes in Electrical Engineering 726,
https://doi.org/10.1007/978-981-33-4110-4_10

In this paper, we have realized a broadband thermo-optic switch with low-power consumption based on a silicon ring-bus-ring-bus (RBRB) structure. The RBRB structure has a transmission passband with reversed Fano line shape. Compared with box-like flat-top lineshape generated by coupled-microrings, reversed Fano lineshape has a character that one sideband is steep, while the other one is gentle [10]. Therefore, the RBRB optical switch can be an effective solution to the spectral distortion and sideband attenuation of conventional Fano switch when it routes optical signals with a high bit rate. Moreover, the sharp Fano lineshape can dramatically reduce the required wavelength shift of the transmission spectrum for switching, which means that the power consumption for switching of the device can be reduced.

2 Device Structure and Theoretical Analysis

As illustrated in Fig. 1, the RBRB structure is composed of a central waveguide sandwiched by two ring resonators, one of which is coupled to another bus waveguide. The widths of ring and bus waveguides are denoted by W_r and W_b, respectively. R_1 (R_2) is the radius of ring 1 (ring 2), and g_1 (g_2) is the gap between ring 1 (ring 2) and the nearby bus waveguide. The input, through, and drop ports are also specified. It is known that Fano resonance is generated from the interference between a continuum state and a discrete state. Therefore, the Fano resonance will appear when the low-Q resonance (continue state) and high-Q resonance (discrete state) are interfered in the RBRB system.

To analyze the RBRB geometry, we established a theoretical model based on the temporal coupled-mode theory (T-CMT). Here, the direct coupling between the two resonators is also considered [11, 12]. We use $1/\tau_{e1}$ ($1/\tau_{e2}$) to describe the coupling efficiency between ring 1 (ring 2) and the central bus (central and drop buses). The direct coupling coefficient between ring 1 and ring 2 is denoted as μ. And $1/\tau_{o1}$ ($1/\tau_{o2}$) and λ_1 (λ_2) are the decay rate due to intrinsic loss and resonant wavelength of ring 1 (ring 2), respectively. Under the steady-state conditions, the transmission of the through (t = S_{thr}/S_{in}) and drop (d = S_{drop}/S_{in}) ports can be derived as

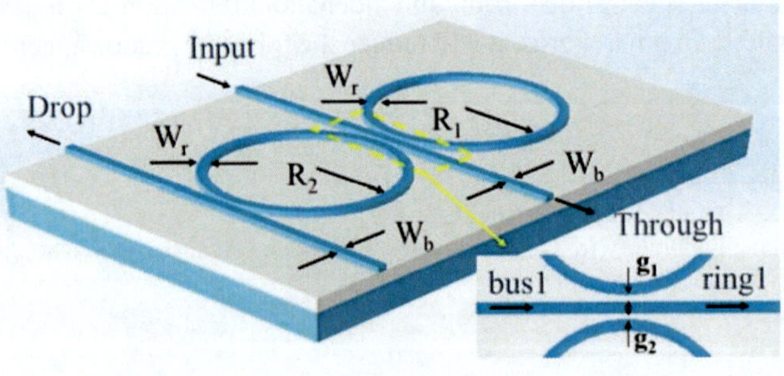

Fig. 1 Schematic of the RBRB system

$$t = 1 + \frac{\gamma_2 \frac{2}{\tau_{e1}} + \gamma_1 \frac{2}{\tau_{e2}} - 4\sqrt{\frac{1}{\tau_{e1}\tau_{e2}}}(i\mu - \sqrt{\frac{1}{\tau_{e1}\tau_{e2}}})}{\gamma_1\gamma_2 - (i\mu - \sqrt{\frac{1}{\tau_{e1}\tau_{e2}}})^2}, \tag{1}$$

$$d = \frac{\gamma_1 \frac{2}{\tau_{e2}} - 2\sqrt{\frac{1}{\tau_{e1}\tau_{e2}}}(i\mu - \sqrt{\frac{1}{\tau_{e1}\tau_{e2}}})}{\gamma_1\gamma_2 - (i\mu - \sqrt{\frac{1}{\tau_{e1}\tau_{e2}}})^2}, \tag{2}$$

where $\gamma_1 = i(2c/\lambda) - 01i(2\pi c/\lambda_1) - 1/\tau_{o1} - 1/\tau_{e1}$, and $\gamma_2 = i(2\pi c/\lambda) - i(2\pi c/\lambda_2) - 1/\tau_{o2} - 1/\tau_{e2} - 1/\tau_{e2}$.

From Eqs. (1) and (2), we have obtained conventional Fano lineshapes and reversed Fano lineshapes in the RBRB system, as shown in Fig. 2a–d and e–h, respectively. With $1/\tau_{o2}$ and $1/\tau_{e2}$ fixed, the lineshapes depend on the values of μ, $1/\tau_{o1}$, and $1/\tau_{e1}$. These parameter values indicate that ring 1 and ring 2 are in the over-coupling regime and under-coupling regime, respectively. It is observed that the reversed Fano lineshapes have a bandwidth much larger than the conventional Fano lineshapes. Hence, the reversed Fano resonance will overcome the limitation of narrow bandwidth of the conventional Fano resonance when it operated as an optical switch.

Then, the RBRB structure is also simulated by two-dimensional finite-difference time-domain (2D-FDTD) method. The structural parameters are set as $R_1 = R_2 = 10\ \mu\text{m}$, $W_r = 0.45\ \mu\text{m}$, $g_1 = 0.10\ \mu\text{m}$, and $g_2 = 0.12\ \mu\text{m}$. The simulation region is bounded by perfectly matched layers. As seen in Fig. 3, left/right-reversed Fano line shapes are generated, and the FDTD results fit the T-CMT results very well.

3 Device Fabrication

Silicon has a large thermo-optic coefficient ($dn/dT = 1.86 \times 10^{-4}\ \text{K}^{-1}$) at the wavelength of 1550 nm. We fabricated a thermo-optical switch based on a RBRB system with a heater in silicon-on-insulator (SOI). The fabricated device is shown in Fig. 4a. All the waveguides are firstly defined on the top-silicon layer with a height of 220 nm, and then covered by a ~1 μm-thick SiO_2 using plasma-enhanced chemical vapor deposition, as seen in Fig. 4b. Then the heater was made up of ~100 nm–thick titanium, and the pads and electric lines were constructed by a ~1000 nm-thick aluminium.

4 Measurement Results and Analysis

As shown in Fig. 5a, reversed Fano lineshapes are observed when a direct current voltage of 2 V is applied on the microheater of ring 1. The experimental results

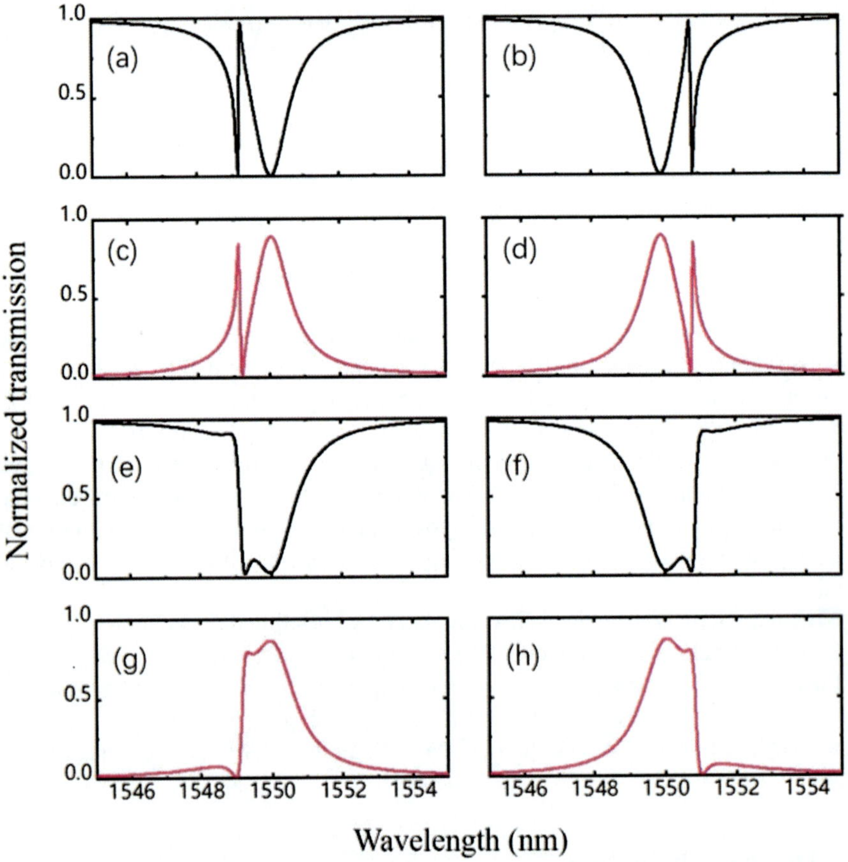

Fig. 2 Four types of Fano lineshapes in the RBRB system. In **a–d**, $1/\tau_{o1} = 1 \times 10^9$ rad/s, $1/\tau_{e1} = 1 \times 10^9$ rad/s. In **e–h**, $1/\tau_{o1} = 1.2 \times 10^{10}$ rad/s, $1/\tau_{e1} = 2 \times 1011$ rad/s. In (**a**, **c**, **f**, **h**), $\mu = -1.8 \times 10^{11}$, and in (**b**, **d**, **e**, **g**), $\mu = 2.1 \times 10^{11}$. In (**a**, **c**, **e**, **g**), $\lambda_1 = 1549.22$ nm, and in (**b**, **d**, **f**, **h**), $\lambda_1 = 1550.78$ nm. They share the same parameter value of $\lambda_2 = 1550.00$ nm, $1/\tau_{o2} = 3 \times 10^{10}$ rad/s, and $1/\tau_{e2} = 2.5 \times 10^{11}$ rad/s. Black solid lines are the transmission spectra at the through port. Magenta solid lines are the transmission spectra at the drop port

(dark line) agree well with the theoretical results (dash line) using T-CMT. The measured spectra show that the reversed Fano lineshapes has a 3-dB bandwidth more than 280 GHz, an extinction ratio of 28 dB. The average slope rate of the right bandedge of Fano resonance is 60.8 dB/nm, while the maximum slope rate exceeds 220 dB/nm. The output spectra can be tuned via changing the voltages on the heater. To characterize the Fano switch, we carried out a simulation and input 40-Gbit/s second-order Gaussian nonreturn-to-zero (NRZ-PRBS) signals for the device. If the right bandedge of the Fano resonance is used for the on/off switching, owing to the large slope rate of revesed Fano lineshape (60.8 dB/nm), the required wavelength shift is only 0.22 nm, corresponding to a power consumption of 1.46 mW. If the left

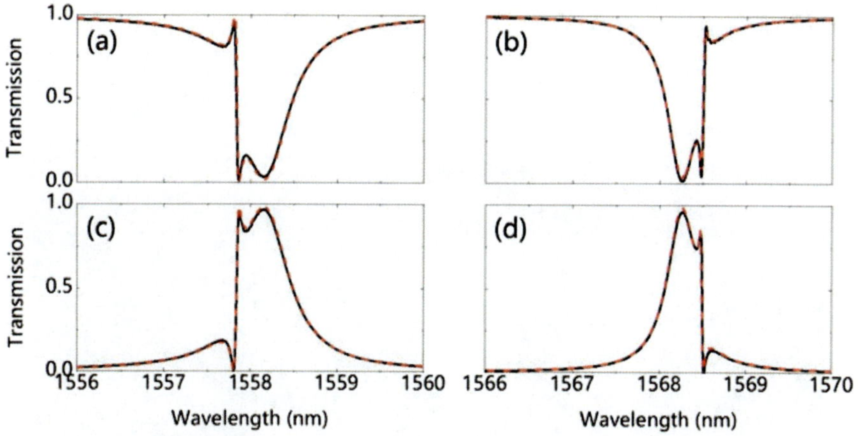

Fig. 3 **a** and **c** left-reversed Fano lineshapes at through port and drop port with $W_b = 0.25\ \mu\text{m}$. **b** and **d** right-reversed Fano lineshapes at through port and drop port with $W_b = 0.45\ \mu\text{m}$. The black solid lines and red dash lines represent the FDTD and T-CMT results, respectively

bandedge (slope rate of 10.6 dB/nm) is used, the required wavelength shift is 2.8 nm, corresponding to a 18.6-mW power consumption.

To reveal the potential data rates in switching, we further performed simulations using 80, 160, 240, and 320 Gbit/s input signals. From the results in Fig. 6, no significant signal distortion effects are observed even for 240 Gbit/s. From Table. 1, it is found that our device has a much larger bandwidth. Because of the sharp bandedge of the reversed Fano resonance, the required wavelength shift for switching is reduced significantly. The crosstalks at the through and drop ports both are as low as −28 dB. Note that the transmission spectrum is nearly flat from 1562 nm at the through port. The insertion losses at the two output ports are only 0.1 dB.

5 Conclusion

In summary, we fabricated a thermo-optic switch based on the RBRB structure in silicon. It is found that the power consumption for switching a 40-Gbit/s signal can be as low as 1.46 mW. Furthermore, we estimated the signal transmission with higher speed data rates (80, 160, 240 and 320 Gbit/s), and the distortion of signals is negligible. Due to the broad bandwidth and high slope rate, the proposed Fano optical switch in silicon will be an effective solution to the next-generation on-chip optical interconnects.

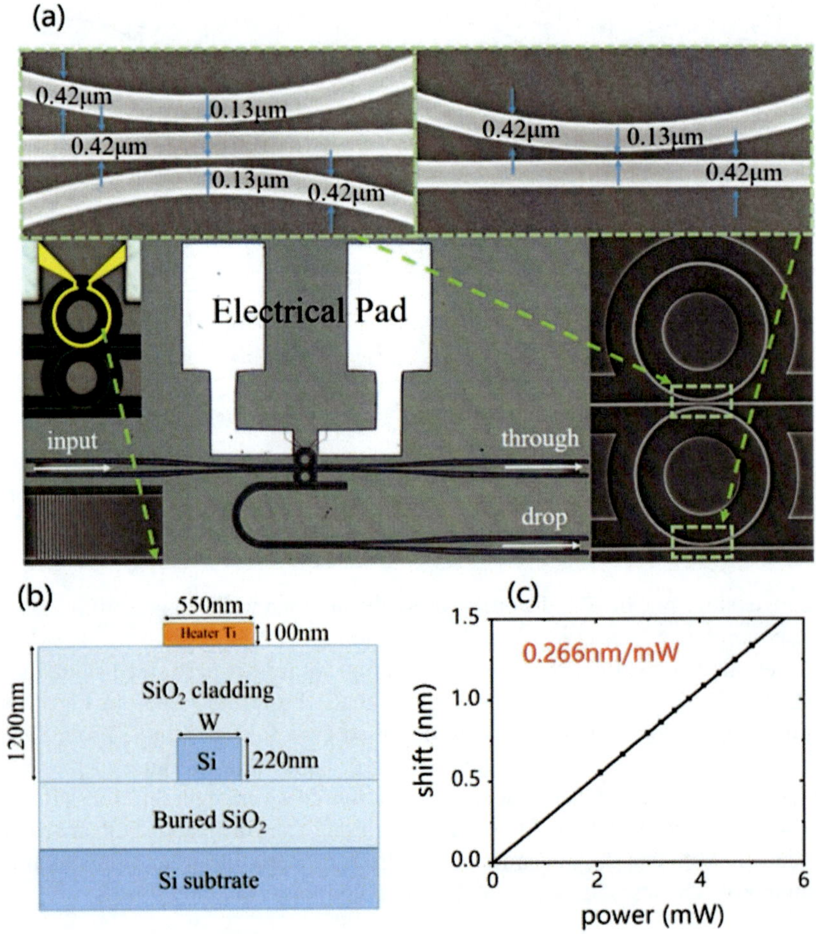

Fig. 4 **a** The images of the fabricated silicon switch based on a RBRB system. **b** The cross-section of the ring waveguide with a top Ti micro heater. **c** Resonance shift as a function of heating power (the dots are measured results). The thermal tuning efficiency is 0.266 nm/mW

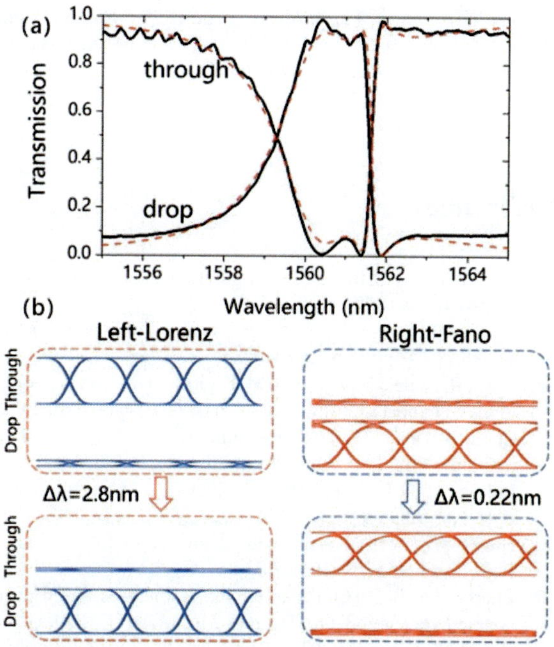

Fig. 5 **a** Normalized transmission spectra of the device. The solid lines and dash lines are the experimental curves and fitting curves, respectively. **b** Eye diagrams of output signals with 40-Gbit/s second-order Gaussian NRZ-PRBS input signals based on simulation

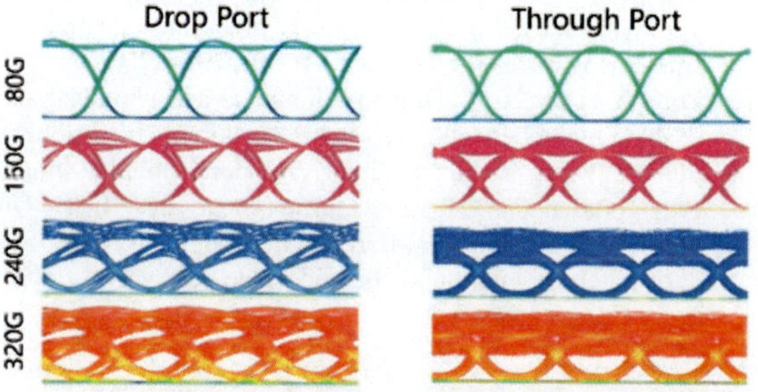

Fig. 6 Eye diagrams with 80/160/240/320-Gbit/s NRZ-PRBS signals at the output ports

Table 1 Comparison between our switch and the previously reported microring switches

Various switch structure	Bandwidth (GHz)	On/off shift (nm)	Thr/drop loss (dB)	Thr/drop crosstalk (dB)
Single microring	90	1.3	0.33/1.64	−11/–
2nd-order coupled-microring	60	2.8	2/0.4	−16.6/−9.8
3nd-order coupled-microring	74	0.54	0.08/1.96	−23/–
10nd-order coupled-microring	80–140	1	3.1–4.9/−4.3–5.9	−10/−50
Our device	280	0.45	0.1/0.1	−28/−28

Acknowledgements This work is supported by National Natural Science Foundation of China (NSFC) (Grants 61675084 and 61775094), and by the Opened Fund of the State Key Laboratory of Integrated Optoelectronics No. IOSKL2018KF14.

References

1. Fano U (1961) Effects of configuration Interaction on intensities and phase shifts. Phys Rev 124(6):1866–1878
2. Yu Y et al Fano resonance control in a photonic crystal structure and its application to ultrafast switching. Appl Phys Lett 105(6)
3. Asadi R, Malek-Mohammad M, Khorasani S (2011) All optical switch based on Fano resonance in metal nanocomposite photonic crystals. Opt Commun 284(8):2230–2235
4. Cheng F et al (2012) Tuning asymmetry parameter of Fano resonance of spoof surface plasmons by modes coupling. Appl Phys Lett 100(13)
5. Luk'yanchuk B et al (2010) The Fano resonance in plasmonic nanostructures and metamaterials. Nat Mater 9(9):707–715
6. Harris SE (1989) Lasers without inversion: Interference of lifetime-broadened resonances. Phys Rev Lett 62(9):1033–1036
7. Zhao G et al (2016) Tunable Fano resonances based on microring resonator with feedback coupled waveguide. Opt Express 24(18):20187–20195
8. Heuck M, Kristensen PT, Elesin Y, Mork J (2013) Improved switching using Fano resonances in photonic crystal structures. Opt Lett 38(14):2466–2468
9. Lee BG et al (2006) Transmission of high-data-rate optical signals through a micrometer-scale silicon ring resonator. Opt Lett 31(18):2701–2703
10. Zhou X et al (2014) On-chip biological and chemical sensing with reversed fano lineshape enabled by embedded microring resonators. IEEE J Sel Top Quantum Electron 20(3):35–44
11. Suh W, Wang Z, Fan S (2004) Temporal coupled-mode theory and the presence of non-orthogonal modes in lossless multimode cavities. IEEE J Quantum Electron 40(10):1511–1518
12. Souza MC, Rezende GF, Barea LA, Wiederhecker GS, Frateschi NC (2016) Modeling quasi-dark states with temporal coupled-mode theory. Opt Express 24(17):18960–18972

Study on the Characteristics of All Fiber Comb Interleaver with Unequal Bandwidth

Bao Ge Zhang, Tian Peng Wang, Jing Jing Tian, and Yao Yao

Abstract An all-fiber comb interleaver with unequal bandwidth is proposed, which is composed of an asymmetrical Mach–Zehnder interferometer and a cascaded 3 × 3 single-mode fiber coupler. The expression of optical output spectrum is obtained by light transmission theory. And the theoretical analysis is carried out, the theoretical analysis shows that when the length differences of two pairs of optical fiber interference arms are equal, and the value of coupling coefficient of each coupler is proper. The proposed all-fiber comb interleaver can be realized a unequal bandwidth output on even and odd channels, through test and calculation, the passband bandwidths of the even and odd channel are greater than 20 Gb/s and 60 Gb/s respectively, and channel isolations are greater than 25 dB, so, the proposed all-fiber comb interleaver can be used for 10 Gb/s + 40 Gb/s communication system. Compared with other comb interleavers with unequal bandwidth, the proposed all-fiber comb interleaver not only has the advantages of wide passband and simple structure, but also is an all fiber comb interleaver with lower cost.

Keywords Fiber Coupler · Mach–Zehnder interferometer · All fiber comb interleaver · Unequal bandwidth

1 Introduction

One of the core devices of dense wavelength division multiplexing (DWDM) system is optical wavelength division multiplexing/demultiplexer, the comb interleaver is a

B. G. Zhang (✉) · J. J. Tian · Y. Yao
School of Automation and Electrical Engeering, Lanzhou Jiaotong University, Gan Su, Lan Zhou 730070, China
e-mail: 276497535@qq.com

T. P. Wang
School of Architecture and Urban Planning, Lanzhou Jiaotong University, Gan Su, Lan Zhou 730070, China

Y. Peng and X. Dong (eds.), *Proceedings of 2019 International Conference on Optoelectronics and Measurement*, Lecture Notes in Electrical Engineering 726,
https://doi.org/10.1007/978-981-33-4110-4_11

new type of photon device. The optical comb interleaver can effectively increase the number of multiplexed channels on optical fiber, at the same time, the technology of the optical comb interleaver is not too complex and the cost is not too high, so, it is a new type of multiplexing/demultiplexing device with rapid development and great prospect. There are many schemes that can realize the optical comb interleaver [1–5], but the bandwidth of parity communication channel of these realized optical comb interleavers are equal to each other respectively, in the current practical system, the coexistence of 10 Gb/s and 40 Gb/s is the development trend of DWDM Technology in recent years. There is an approximate relationship between the optimal bandwidth Δf and the transmission rate v. For the systems with transmission rates of 10 Gb/s and 40 Gb/s, the odd channel bandwidth and the even channel bandwidth shall be 15 GHz and 60 GHz respectively. Therefore, in the process of upgrading from 10 to 10 Gb/s + 40 Gb/s, the interleaver with unequal bandwidth as a multiplexing demultiplexer has greater flexibility. At the same time, the interleaver with unequal bandwidth can has a higher advantage in improving the bandwidth utilization rate of the communication system, and the interleaver with unequal bandwidth can reduce the cost of the communication system. The reported optical comb interleavers with unequal width of different kinds are: Michelson interferometer type, birefringent fiber loop mirror, birefringent gires tournois type, coupler cascade type [6–8]. However, the interleavers structure with unequal bandwidth provided in these literatures have many problems, such as large insertion loss, complex structure and difficult fabrication.

An all-fiber comb interleaver with unequal bandwidth is proposed, which is composed of an asymmetrical Mach–Zehnder interferometer and a cascaded 3 × 3 single-mode fiber coupler in the paper, which can realize the unequal bandwidth output for being used in the communication system with 10 Gb/s + 40 Gb/s. In terms of technology and structure, the proposed interleaver has low cost and simple fabrication. When the 3 × 3 coupler is drawn, the free port of the 3 × 3 coupler is used as the monitoring end to avoid the interference effect of the monitoring light. The coupling coefficient of each coupler can be accurately controlled. At the same time, the structure of the proposed interleaver is not sensitive to the coupling coefficient of the coupler, which reduces the manufacturing difficulty of the proposed interleaver, and overcomes the shortcomings of the previous production of optical comb interleaver with unequal bandwidth, and has a certain practical value.

2 Interleaver Structure and Theoretical Analysis

The structure of the proposed all fiber comb interleaver is composed of two 1-shaped 3 × 3 fiber couplers (C_0 and C_1) and one 2 × 2 fiber coupler (C_2) in Fig. 1. In Fig. 1, an asymmetrical Mach–Zehnder interferometer is composed of couplers C_1 and C_2, C_0 and C_1 are connected by a pair of optical fiber interference arms l_1 and l_2, C_1 and C_2 are connected by another pair of optical fiber interference arms l_3 and l_4.

Assuming that the light is only input from port 1 of the coupler C_0, when the input light E^{in} (1,0,0) is input from input port 1, after passing through the optical

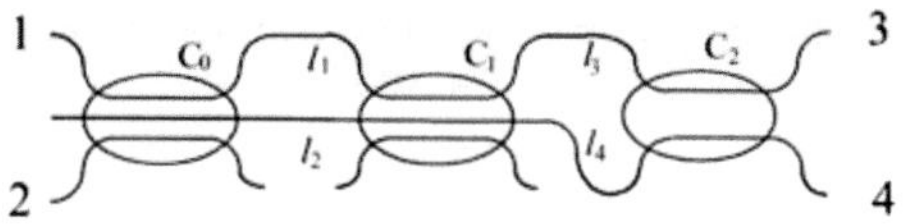

Fig. 1 The structure of optical comb interleaver with unequal bandwidth

fiber coupler C_0, the output light is allocated into the interference arms l_1 and l_2, and is transmitted to the coupler C_1, the light is output after the first interferencing, and is allocated into the interference arms l_3 and l_4, then is again transmitted to the coupler C_2, and is output finally after the second interferencing, then the light is finally output from port 3 and port 4, the final output light field is recorded as F.

Neglecting the fiber transmission loss and the insertion loss of the fiber coupler, θ_1 and θ_2 represent the transmission phase of the two pairs of interference arms l_1 and l_2, l_3 and l_4 respectively, and they are satisfied the corresponding expressions, that is, $\theta_1 = \beta \Delta l_1 = \beta(l_2 - l_1) = 2\pi n_{eff} \Delta l_1/\lambda$, $\theta_2 = \beta \Delta l_2 = \beta(l_4 - l_3) = 2\pi n_{eff} \Delta l_2/\lambda$, n_{eff} is the effective refractive index of the conductive mode of the fiber, and β is the propagation constant of the fundamental mode in the fiber, K_n ($n = 0, 1, 2$) is the coupling coefficient of each coupler. At the output end of the proposed interleaver, the output light intensity is recorded as P, the light intensity of port 3 and port 4 are $P_3 = |\boldsymbol{F}_3|^2$, $P_4 = |\boldsymbol{F}_4|^2$, respectively. In the design of all fiber optical comb interleaver with unequal bandwidth, one basic transmission matrixes of optical transmission is scattering matrixes of the couplers, which are $\boldsymbol{M}_r$ ($r = 0, 1$) and $\boldsymbol{M}_q$, and the other basic transmission matrix of optical transmission is phase delay matrixes of the relative arm length difference, which are $\boldsymbol{L}_1$ and $\boldsymbol{L}_2$.

$$\mathbf{M}_r = \begin{bmatrix} 0.5(1+\cos k_r) & j0.5\sqrt{2}\sin k_r & 0.5(\cos k_r - 1) \\ j0.5\sqrt{2}\sin k_r & \cos k_r & j0.5\sqrt{2}\sin k_r \\ 0.5(\cos k_r - 1) & j0.5\sqrt{2}\sin k_r & 0.5(1+\cos k_r) \end{bmatrix}$$

$$\mathbf{M}_q = \begin{bmatrix} \cos k_q & -j\sin k_q \\ -j\sin k_q & \cos k_q \end{bmatrix}$$

$$L_i = \begin{bmatrix} \exp(-j\beta\Delta l_i) & 0 \\ 0 & 1 \end{bmatrix}$$

Here, $r = 0, 1$; $q = 2$; $i = 1, 2$.

According to the theory of the optical coupling transmission, the expression of the output light intensity of the proposed interleaver can be obtained from the scattering matrixes ($\boldsymbol{M}_0$, $\boldsymbol{M}_1$ and $\boldsymbol{M}_2$) and the phase delay matrixes ($\boldsymbol{L}_1$, $\boldsymbol{L}_2$), the output light intensity recorded as P.

$$\boldsymbol{F} = \begin{bmatrix} \boldsymbol{F}_3 \\ \boldsymbol{F}_4 \end{bmatrix} = \boldsymbol{M}_2 \boldsymbol{L}_2 \boldsymbol{M}_1 \boldsymbol{L}_1 \boldsymbol{M}_0 \boldsymbol{E}^{in} \tag{1}$$

The light intensity of port 3 and port 4 are $P_3 = |F_3|^2$, $P_4 = |F_4|^2$. After complicated calculation, P is as follows:

$$P = \begin{bmatrix} P_3 \\ P_4 \end{bmatrix} = A \cdot \theta \tag{2}$$

Here,

$$A = \begin{bmatrix} a_4 & a_3 & a_2 & a_1 & a_0 \\ b_4 & b_3 & b_2 & b_1 & b_0 \end{bmatrix}, \quad \theta = \begin{bmatrix} \cos\theta_1 & \cos\theta_2 & \cos(\theta_2 - \theta_1) & \cos(\theta_2 + \theta_1) & 1 \end{bmatrix}^T$$

A and θ respectively represent the matrix formed by the optical transmission matrix in the transmission process, each element (a_i, b_i, $i = 0, 1, 2, 3, 4$) in A is a function of the coupling coefficient K_n ($n = 0, 1, 2$), each element in θ is a function of the phase difference (θ_1, θ_2).

It can be seen from formula (2) that the output light intensity only includes θ_1, θ_2, $\theta_1 - \theta_2$, $\theta_1 + \theta_2$. These transmission phase difference (caused by arm length difference) of the optical fiber segment is the main factor that can determine the transmission bandwidth of the optical comb interleaver, therefore, the optical comb interleaver with unequal bandwidth can be realized by selecting the appropriate coupling coefficients of the couplers and the arm length difference.

3 Characteristic Analysis

The transmission phase θ_1 is selected to be equal the transmission phase θ_2, that is $\theta_1 = \theta_2 = \theta$, and fiber arm length difference is $\Delta l_1 = \Delta l_2 = 2$ mm, the effective refractive index n_{eff} of optical fiber is 1.457. Then selecting the appropriate coupling coefficient K_n ($n = 0, 1, 2$) of the couplers, the coefficient matrix A needed for the ideal output spectral of the optical comb interleaver can be obtained. Through optimization calculation, $K_0 = \pi/9$, $K_1 = 5\pi/19$ and $K_2 = 10\pi/51$ are selected, so the coefficient matrix A is recorded as A_1, A_1 is as follows:

$$A_1 = \begin{bmatrix} a_4 & a_3 & a_2 & a_1 & a_0 \\ b_4 & b_3 & b_2 & b_1 & b_0 \end{bmatrix} = \begin{bmatrix} -0.0812 & -0.3676 & 0.0599 & -0.1257 & 0.5452 \\ 0.0418 & 0.3676 & -0.0599 & 0.1257 & 0.4136 \end{bmatrix}$$

The calculated output spectrum changing with the wavelength value is as shown in Fig. 2.

In Fig. 2, the solid line represents the normalized output optical power curve of port 3, and the dotted line represents the normalized output optical power curve of port 4. The output spectral clearly shows that the output spectra of port 3 and port

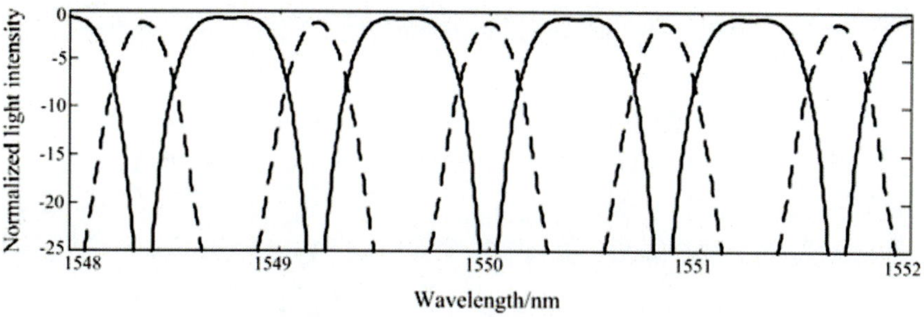

Fig. 2 The output spectrum of the interleaver

4 of the proposed interleaver have good even and odd symmetry. The spectral in the wavelength domain (frequency domain) is composed of a series of transmission peaks with unequal spacing, and the edge of the passband is steep, which is more suitable waveform for the optical interleaver, so the proposed interleaver can be used as an optical comb interleaver. It can also be seen from Fig. 2 that the channel interval is 0.8 nm, the center wavelength complies with ITU regulations. By calculation, the 3 dB bandwidth of port 3's output spectral is greater than 60 GHz, which can be used for 40 Gb/s transmission rate, and the 3 dB bandwidth of port 4's output spectral is greater than 20 GHz, which can be used for 10 Gb/s transmission rate, then the output spectral of the proposed optical comb interleaver is realized unequal bandwidth, so the transmission rate can be 50 Gb/s within 100 GHz bandwidth. At the same time, Fig. 2 clearly shows that the channel isolation degree is greater than 25 dB, which can meet the commercial demand, and the proposed optical comb interleaver has high consistency in the whole working frequency band, it can be seen that the one port of the 3×3 coupler is not utilized from the structure, so the additional loss and insertion loss of the proposed optical comb interleaver are relatively increased, and the normalized output peak value is slightly less than 1.

Owing to the process conditions and other reasons in the making, there is often a deviation between the measured value and the expected value of coupling coefficient of the coupler, the control of the deviation of each parameter is very important in the development of the device [9, 10]. Therefore, in order to study the influence of the deviation on the transmission characteristics of the interleaver, three groups of datas are selected for calculating. When $K_0 = \pi/8$, $K_1 = 5\pi/18$, $K_2 = 10\pi/51$, the coefficient matrix A is recorded as A_2, when $K_0 = 5\pi/37$, $K_1 = 2\pi/7$, $K_2 = 5\pi/24$, the coefficient matrix A is recorded as A_3, when $K_0 = 10\pi/81$, $K_1 = \pi/4$, $K_2 = 5\pi/27$, the coefficient matrix A is recorded as A_4. Then after calculation, A_2, A_3 and A_4 is as follows:

$$A_2 = \begin{bmatrix} a_4 & a_3 & a_2 & a_1 & a_0 \\ b_4 & b_3 & b_2 & b_1 & b_0 \end{bmatrix} = \begin{bmatrix} -0.0938 & -0.3643 & 0.0720 & -0.1296 & 0.5309 \\ 0.0434 & 0.3643 & -0.0720 & 0.1296 & 0.4166 \end{bmatrix}$$

$$A_3 = \begin{bmatrix} a_4 & a_3 & a_2 & a_1 & a_0 \\ b_4 & b_3 & b_2 & b_1 & b_0 \end{bmatrix}$$
$$= \begin{bmatrix} -0.0861 & -0.3676 & 0.0822 & -0.1361 & 0.5107 \\ 0.0282 & 0.3676 & -0.0822 & 0.1361 & 0.4290 \end{bmatrix}$$

$$A_4 = \begin{bmatrix} a_4 & a_3 & a_2 & a_1 & a_0 \\ b_4 & b_3 & b_2 & b_1 & b_0 \end{bmatrix}$$
$$= \begin{bmatrix} -0.0984 & -0.3401 & 0.0591 & -0.1427 & 0.5648 \\ 0.0607 & 0.3401 & -0.0591 & 0.1427 & 0.3961 \end{bmatrix}$$

The simulation results are shown in Fig. 3a–c.

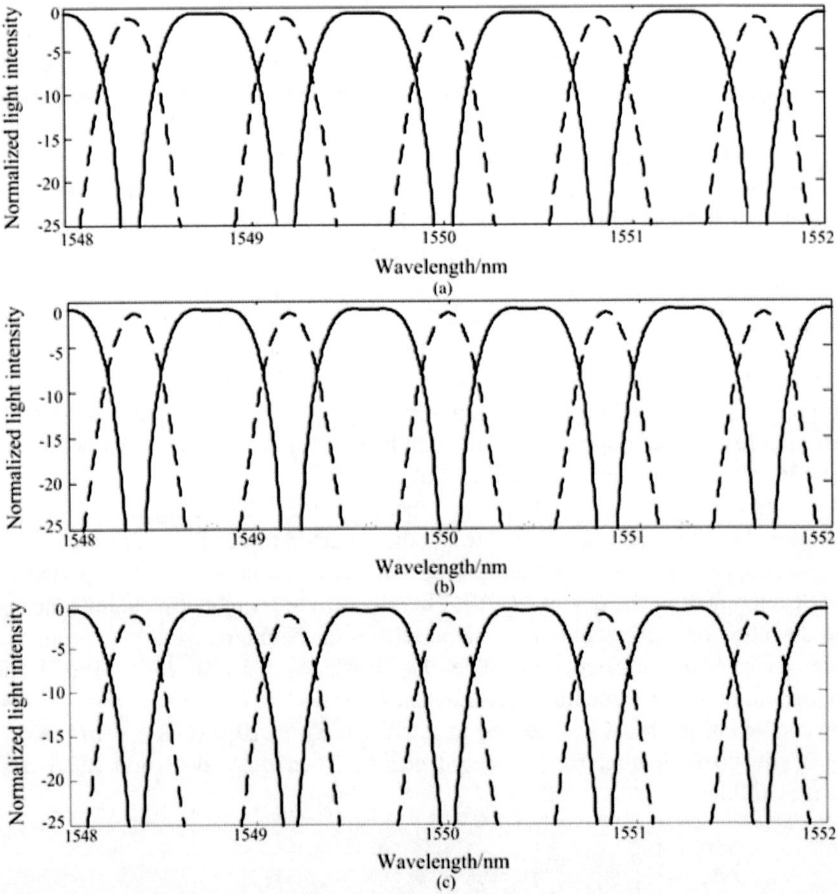

Fig. 3 The different output spectrum of the interleaver. **a** $K_0 = \pi/8$, $K_1 = 5\pi/18$, $K_2 = 10\pi/51$, **b** $K_0 = 5\pi/37$, $K_1 = 2\pi/7$, $K_2 = 5\pi/24$, **c** $K_0 = 10\pi/81$, $K_1 = \pi/4$, $K_2 = 5\pi/27$

Figure 3 shows that although the K_n ($n = 0, 1, 2$) in (a), (b) and (c) are slightly changed, it can be seen that the shape of the output spectral of the two output terminals has little change (almost no change). Therefore, when the coupling coefficients of the three couplers deviate the expected values of coupling coefficient, there is no great influence on the output spectrum of the device. That is to say, if the coupling coefficients of the couplers deviate slightly from the best values, an ideal output spectrum can also be obtained, that is, the requirements of the coupling coefficients of the device are not very strict, which reduces the practical production difficulty of the device.

4 Conclusion

In this paper, a new type of all fiber optical comb interleaver with unequal bandwidth is proposed. It is composed of two 3 × 3 couplers and a 2 × 2 coupler. When the optical coefficients of the coupler and the optical fiber interference arm meet certain conditions, the function of the proposed optical comb interleaver with unequal bandwidth can be realized. The output characteristics of the interleaver are simulated in detail. This proposed optical comb interleaver can be used for simultaneous transmission of 40 and 10 Gb/s, so the transmission rate of 50 Gb/s can be realized in the bandwidth of 100 GHz, so the proposed optical comb interleaver can make full use of the effective bandwidth, and the channel isolation degree is greater than 25 dB, and the channel interval is 0.8 nm. Moreover, the device has good channel uniformity, bandwidth, high consistency and low price. So, the value of the proposed optical comb interleaver is that it can provide an idea for all fiber devices with unequal bandwidth.

Acknowledgements The support of the National Natural Science Foundation of China (No. 61741508) is gratefully acknowledged.

References

1. Kumar N, Shenoy MR, Pal BP (2008) Flattop all fiber wavelength interleaver for DWDM transmission: design analysis, parameter optimization, fabrication and characterization recipe. Optics Commun 281(20):5156–5164
2. Ye X, Zhang M, Ye P (2006) Flat-top interleavers with chromatic dispersion compensator based on phase dispersive free space Mach-Zehnder interferometer. Optics Commun 257(2):255–260
3. Chan W, Chen K, Chan H, Pal BP, Varshney RK (2009) A flattop PLC polymer waveguide interleaver based on folded two-stage-cascaded Y-junction Mach-Zehnder interferometers. Optics Commun 282(5):883–886
4. Pu HL, Lu HW (2015) Design and experiment study of all fiber MZI interleaver based on "8"-shaped fiber ring resonator. J Optoelectronics·Laser 26(2):239–244
5. Lu Y, Wei X, Gao M, Lu H, Luo G (2018) Theoretical analysis of a novel all fiber interleaver by using improved Mach-Zehnder interferometer. J Optoelectron Laser 29(6):582–587

6. Zhang T, Kai C, Zhao S (2005) A novel unequal passband interleaver employing high birefringence fibersagnac loop mirro. J Optoelectronics·Laser 16(4):436–439
7. Zhang B, Huang D (2003) Characteristics study on birefrigent gires-tournois interleaver. Acta Optica Sinic 23(9):1068–1070
8. Zhang RF, Wang SH, Ge CF (2006) Design of different bandwidth interleaver. J Tianjin Univ 39(3):365–368
9. Pan W, Zou L, Luo B (2005) Theoretical study of all optical XOR using Mach-Zehnder interferometer. Opt Precision Eng 13(3):339–347
10. Shuai C, Duan J, Zhong J (2005) Technical sensitiveness in the rheological manufacture progress of fused tapercoupler. Opt Precision Eng 13(1):40–46

Two-Dimensional Beam Steering in Optical Phased Array with Grating Array Superlattices

Huihui Zhang, Qiang Liu, Qiqi Yuan, and Qingzhong Huang

Abstract In this paper, we designed and fabricated an optical phased array with a narrow antenna spacing in silicon on insulator. The crosstalk is reduced substantially with a spacing of only 0.9 μm between antennas using superlattice grating waveguides. A transversal 24.2° beam steering range was observed in the far-field of view by tuning the voltages on each channel. Meanwhile, by sweeping the wavelength from 1540 to 1570 nm, the main beam can scan in a range of 6.45° in the longitudinal direction.

Keywords Subwavelength structures · Integrated optics devices · Integrated optoelectronic circuits · Diffraction gratings nology

1 Introduction

Optical phased array (OPA), as a kind of all-solid state light steering device, is playing a significant role in many applications, such as light detection and ranging (LiDAR), free-space communications and optical imaging [1, 2]. So far, most demonstrated OPAs have a relatively large grating spacing (typically >2 μm), resulting in a small steering range in the far field [3–5]. Wide steering angle can be achieved using even narrow spacing. However, the crosstalk between antennas would become significant, which is detrimental for the beam quality [6].

In this paper, we designed and fabricated a 1 × 10 superlattice OPA with a grating pitch of 0.9 μm in silicon on insulator (SOI). From our simulations, the beam is able to steer in a two-dimensional range of 106° × 32.3° through thermo-optic effect and wavelength sweeping from 1400 to 1600 nm. In experiment, a 22.8° beam steering range in the far-field of view was observed by applying different voltages on each

H. Zhang · Q. Liu · Q. Yuan · Q. Huang (✉)
Wuhan National Laboratory for Optoelectronics, Huazhong University of Science and Technology, 430074 Wuhan, China
e-mail: huangqz@mail.hust.edu.cn

Y. Peng and X. Dong (eds.), *Proceedings of 2019 International Conference on Optoelectronics and Measurement*, Lecture Notes in Electrical Engineering 726,
https://doi.org/10.1007/978-981-33-4110-4_12

channel. The steering range is limited by the measurement system. By sweeping the wavelength from 1540 to 1570 nm, a range of 7.1° is achieved in longitudinal direction. The full width at half maximum (FWHM) of the main beam in the far field is 10.9° × 5.1°, roughly consistent with 8.9° × 3.7° in the simulation results.

2 Structure and Principle

As shown in Fig. 1, the device structure is based on a SOI wafer with a 2-μm-thick buried oxide layer and a 220-nm-thick top-silicon layer. The light is firstly coupled into the on-chip waveguides by a grating coupler from a tunable laser. Then it is split into 10 channels (or array waveguides) through a star coupler, and each channel has the same phase in the beginning. Then, each channel has an optical phase shifter tuned by a heater.

The emitting region is displayed in detail in Fig. 2. Five shallow-etched grating antennas with different widths and periods construct a sub-array, which are separated by 0.9 μm. By tuning the phase of each channel, the beam can steer in the phased-array axis (Ψ, in yellow plane), while it is steered in another axis (θ, in green plane) under control of wavelength. The steering angle in two dimensional directions can be written as

$$\sin\psi = \frac{\lambda\Delta\varphi}{2\pi d} \tag{1}$$

$$\sin\theta = \frac{\Lambda_{gr} n_{eff,gr} - \lambda}{n_0 \Lambda_{gr}} \tag{2}$$

where λ is the operating wavelength in free space, d is the grating pitch, $\Delta\varphi$ is the phase difference from one element to the next, Λ_{gr} is the grating period, $n_{eff,gr}$ is the effective refractive index of Bloch mode in sub-wavelength gratings, and n_0

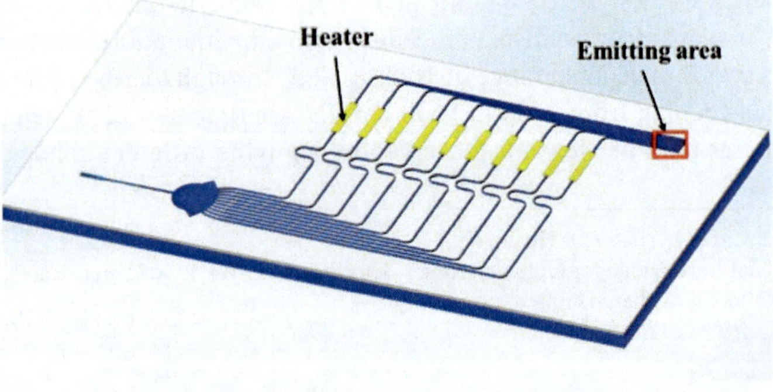

Fig. 1 3-D schematic view of the OPA with 10 elements

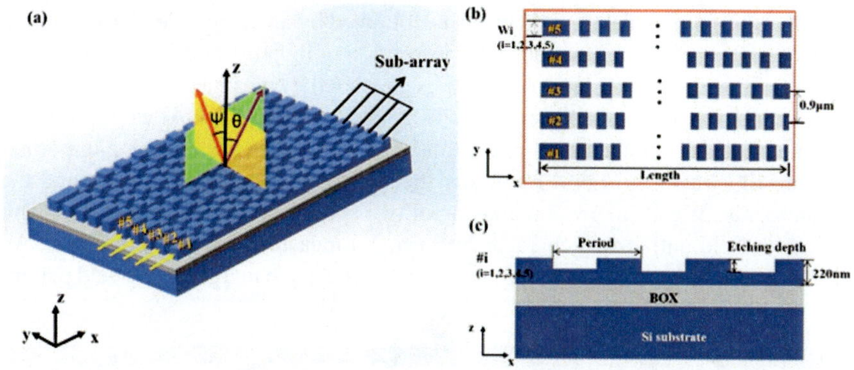

Fig. 2 **a** Emitting area of OPA with 10 elements. **b** Schematic diagram of the grating sub-array. **c** Cross-sectional view of single grating

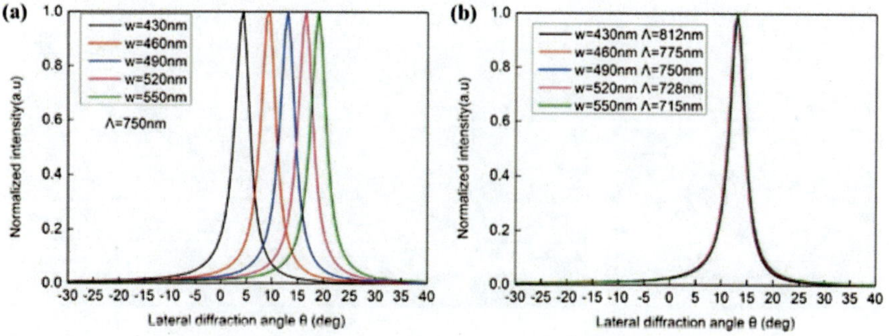

Fig. 3 Far-field distributions of the gratings in sub-array with the same period of 750 nm (**a**) and adjusted periods for the same radiation angle (**b**)

is the refractive index of background that is defined as ($n_0 = 1$). The grating widths in sub-array are set as 550 nm, 490 nm, 430 nm, 520 nm and 460 nm, respectively. To get the same diffraction angle in the far field of view (FOV), their periods are adjusted individually according to Eq. (2). Figures 3a, b show the far-field diffraction angles of five grating antennas with the same and different periods, respectively, at a wavelength of 1500 nm.

3 Device Fabrication and Measurement

According to the design, we fabricated SOI based OPAs. Firstly, the waveguides and star coupler were defined by e-beam lithography (EBL), and they were fully etched on top-silicon layer through dry etching, which can minimize the optical coupling and eliminate the phase perturbations induced by etching depth variations [7]. Then the teeth of gratings were determined by EBL, followed by an etching depth of 70 nm to form the grating emitters. To reduce the absorption of light by the metal electrode, an oxide cladding layer with depth of ~1 μm is deposited by plasma enhanced chemical vapor deposition. Finally, 100-nm-thick Ti metal and 1-μm-thick Al were deposited respectively to form the thermal phase tuner. The whole device and some important parts are illustrated in Fig. 4. Large distance between the grating coupler and the antenna arrays is used to prevent aliasing of two spots within FOV in measurement.

The far-field output was characterized using a measurement setup in Fig. 5. An objective lens with a numerical aperture of 0.3 was used to image the far field pattern

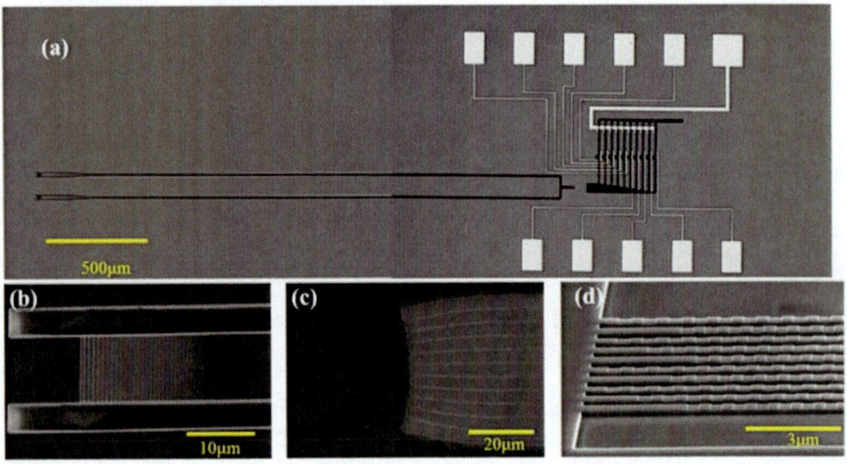

Fig. 4 **a** Top-view microscope image of the 1 × 10 OPA; **b** SEM of the grating coupler at the input port; **c** SEM of the star coupler; **d** SEM of the emitting area

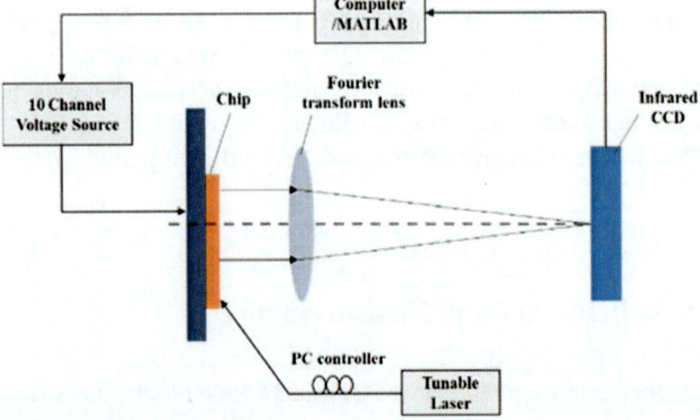

Fig. 5 Schematic diagram of measurement setup for far-field pattern imaging

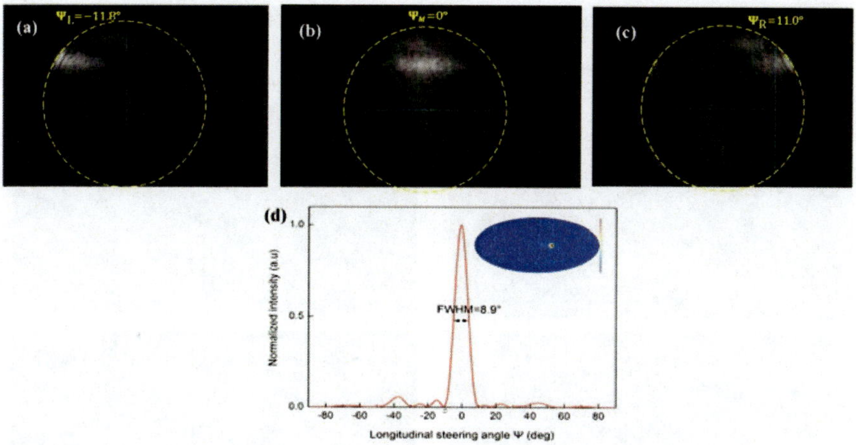

Fig. 6 **a–c** Measured far-field images at different phase gradient. **d** Simulated results of 1 × 10 OPA when phased difference between adjacent antennas is set as 0

into the Fourier plane, and the final image was projected onto an infrared camera. We mounted the device on a printed circuit board (PCB) by an ultraviolet epoxy. All the thermal tuners were wire-bonded to PCB for applying voltages easily. The initial far-field pattern has unwanted sidelobes due to the random phase error in each channel. So we applied an algorithm to control voltages independently and correct the phase errors.

According to Eq. (1), the main beam is steered in the transversal direction (angle Ψ) under different $\Delta\varphi$. Then, we applied different voltages for a fixed operation wavelength of 1550 nm, and the measured images are shown in the Fig. 6a–c.

Here, the observable field range is ±17.4°, which is limited by the objective lens. In experimental measurement, we captured a ~22.8° steering range from −11.8° to 11.0° in FOV by applying different voltages on each channel, and the FWHM of the spot at the middle part is 13.8°. It should be noted that the steering range in the transversal direction is limited by our measurement system. According to the simulation, a 106° steering range can be achieved when the phased difference is set as 17π/18 between adjacent antennas. The simulated data with the same phase difference are shown in Fig. 6d.

In the longitudinal direction, the main beam is scanned by tuning the operation wavelength. In our experiment, the wavelength varies from 1540 to 1570 nm, and the measured intensity patterns are displayed in Fig. 7. The spot is mostly concentrated at 1550 nm in Fig. 7c, while it appears divergent when the wavelength moves to 1570 nm as illustrated in Fig. 7d. That is because the diffraction pattern of each grating in sub-array changes slightly in the far field when the operation wavelength changes, and this situation can be improved by further optimizing the algorithm.

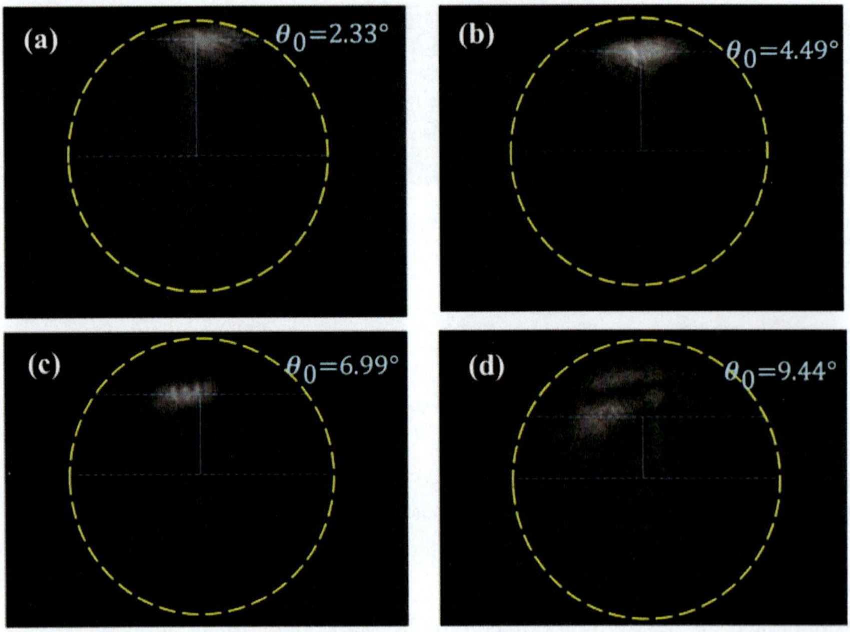

Fig. 7 Measured far-field images at a wavelength of **a** 1540 nm, **b** 1550 nm, **c** 1560 nm, and **d** 1570 nm

4 Conclusion

We designed and fabricated a 1 × 10 optical phased array based on the superlattice structure, and the grating pitch can be reduced to 0.9 μm with a low crosstalk. In our experiment, a 22.8° steering range from −11.8° to 11.0° in the transversal direction is observed by adjusting voltages on each channel. And in the longitudinal direction, the wavelength sweeping in a 40 nm range provides ~7° beam steering range in the far-field. The FWHM of the main beam is about 8.9° × 3.7°, which can be reduced by enlarging the emitting area of OPA and optimizing the phase error correction.

Acknowledgements This work is supported by National Natural Science Foundation of China (NSFC) (Grants 61675084 and 61775094), and by the Opened Fund of the State Key Laboratory of Integrated Optoelectronics No. IOSKL2018KF14.

References

1. Sun J, Timurdogan E, Yaacobi A et al (2013) Large-scale nanophotonic phased array. Nature 493(7431):195–199

2. Hutchison DN, Sun J, Doylend JK et al (2016) High-resolution aliasing-free optical beam steering. Optica 3(8):887–890
3. Doylend JK, Heck MJR, Bovington JT et al (2012) Hybrid III/V silicon photonic source with integrated 1D free-space beam steering. Opt Lett 37(20):4257–4259
4. Karel VA et al (2011) One-dimensional off-chip beam steering and shaping using optical phased arrays on silicon-on-insulator. J Lightwave Technol 29(23):3500–3505
5. Gabrielli LH, Hernandez-Figueroa HE (2015) Aperiodic antenna array for secondary lobe suppression. IEEE Photonics Technol Lett 28(2):209–212
6. Song W, Gatdula R, Abbaslou S et al (2015) High-density waveguide superlattices with low crosstalk. Nature Communications 6:7027
7. Zhang Y, Ling YC, Zhang KQ et al (2019) Sub-wavelength-pitch silicon-photonic optical phased array for large field-of-regard coherent optical beam steering. Opt Express 27(3):1929–1940

Effects of Source/Drain Electrodes on the Performance of Organic Phototransistors Based on Cooper Phthalocyanine

Zhuoli Zhou, Chao Han, and Yingquan Peng

Abstract Optimizing the materials of the source/drain electrode-layer of organic phototransistors (OPTs) is very important in order to achieve high device performance. In this work, OPTs based on cooper phthalocyanine (CuPc) with different metal materials electrodes were fabricated. These devices show differences in performance. The device with gold electrodes shows a high photoresponsivity of 14.5×10^{-2} A/W and a high mobility of 102.4×10^{-3} $cm^2/V{\cdot}s$ under illumination, while the copper electrodes device is 1.3×10^{-2} A/W and 4.4×10^{-3} $cm^2/V{\cdot}s$ for the value. But, the copper device shows better on maximum photo/dark current ratio of 488.5 than the gold device of 59.9.

Keywords Organic phototransistors (OPTs) · Metal electrodes

1 Introduction

Organic field-effect transistors (OFETs) have attracted more and more attention because of their advantages, such as low cost fabrication, light weight and good flexibility, and wide span of applications [1–4]. OFETs that strongly respond to light (i.e., organic phototransistors or OPTs) have an additional control terminal and can be used in photoelectronic circuits as light sensors, photo-switches or signal amplifiers. In the last several years, the performance of OPTs has been improved significantly by developing promising functional materials and creating optimized device structures [5].

But, the intrinsic electrical characteristic of the organic semiconductors is just one of the factors affecting the device performance. At the same time, the contact surface between the organic film and the source/drain (S/D) electrodes is also crucial, which

Z. Zhou · C. Han · Y. Peng (✉)
College of Optical and Electronic Technology, China Jiliang University, Hangzhou, People's Republic of China
e-mail: yqpeng@cjlu.edu.cn

Y. Peng and X. Dong (eds.), *Proceedings of 2019 International Conference on Optoelectronics and Measurement*, Lecture Notes in Electrical Engineering 726,
https://doi.org/10.1007/978-981-33-4110-4_13

restricts the development of OPTs, since an inevitable injection barrier always exists between the organic semiconductor and S/D electrodes. However, most research on OPTs has focused mainly on the materials of organic functional layer and different structures, while the contact of the electrodes/organic in the OPTs has received less attention. Therefore, the effect of contact surface on the performance of OPTs maybe can be studied by changing electrode materials.

In this work, a cooper phthalocyanine (CuPc) film is used as the organic functional layer of the OPTs. Meanwhile, the top S/D electrodes are made of four metal (gold, silver, aluminum and copper), respectively. The effects on the performance are analyzed by comparing the photoresponsivity, threshold voltage, mobility, etc. of these four devices.

2 Experiment

As shown in Fig. 1, the device structure of OPTs used in this study is glass/ITO/polyvinyl alcoho (PVA) (710 μm)/CuPc (50 nm)/top metal electrodes (150 nm). The ITO coated glass was etched with 4 mm wide stripes and used as gate electrodes. 0.25 g/ml PVA solution was spin-coated onto the ITO-glass substrate and baked in a vacuum drying box at 200 °C for an hour to be used as the gate insulator. The PVA layer thickness was measured to be 710 μm. The 50 nm thick CuPc layer was deposited evenly under a vacuum of less than 2×10^{-4} Pa at the rate of 0.2 Å/s as the organic functional layer. Finally, the S/D electrodes of different metal materials (Au, Ag, Al and Cu) were thermally evaporated on the CuPc film and patterned with a mask. The length and the width of the channel are 80 μm and 2 mm.

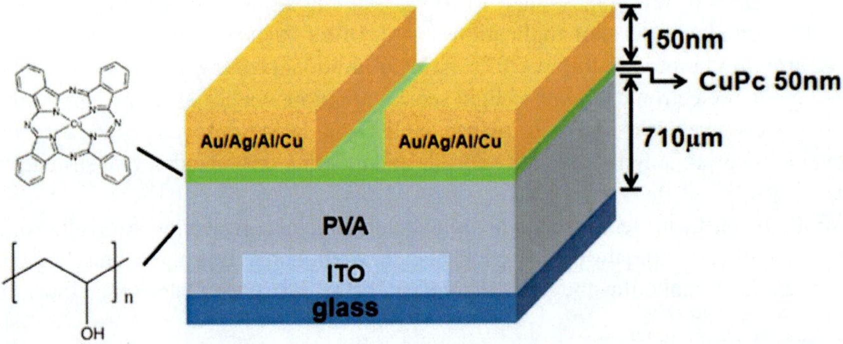

Fig. 1 The structure of OPTs based on Au/Ag/Al/Cu source/drain top electrodes

3 Results and Discussion

We measured the output and transfer characteristics of the four devices under different illumination. The photocurrent I_{photo} can be calculated from the following equation:

$$I_{photo} = |I_{ill} - I_{dark}| \tag{1}$$

where I_{ill} and I_{dark} is the output current in the dark and under illumination, respectively. Figure 2 shows I_{photo} of these devices under illumination with the incident light power of 4.83 mW/cm^2. It can be seen that I_{photo} of the device with gold electrodes almost three times more than other devices.

The optical properties of the photoresponsitivity (*R*) and external quantum efficiency (*EQE*) of the OPTs were extracted and calculated from Eqs. (2) and (3). The threshold voltage (V_{th}) can obtained from the transfer characteristic curve, its value is the intercept on the x-axis of the tangent of the $I_{DS}{}^{1/2}$-V_{GS} curve. Meanwhile, the mobility of the devices can be calculated from Eq. (4) by V_{th}.

$$R = \frac{I_{photo}}{P_{opt}} = \frac{I_{photo}}{P_{\text{int}} \times W \times L} \tag{2}$$

$$EQE = \frac{hc}{q\lambda} R \tag{3}$$

$$I_{DS} = \frac{W}{2L} C_{ox} \mu (V_{GS} - V_{th})^2 \tag{4}$$

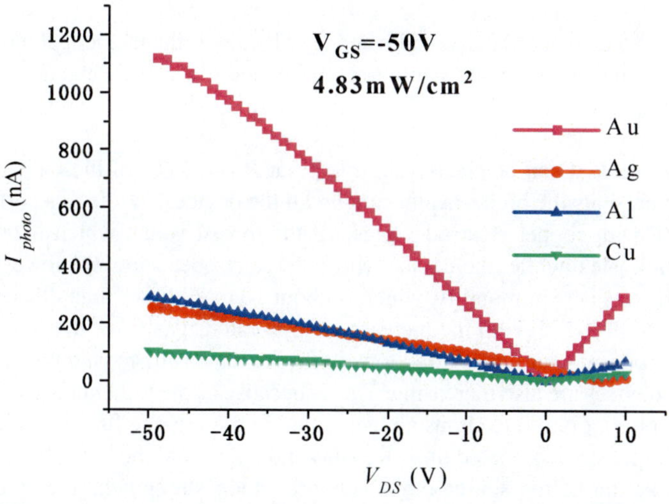

Fig. 2 The I_{photo} as a function of V_{DS} recorded at $V_{GS} = -50$ V under the illumination of $\lambda =$ 650 nm and $P_{int} = 4.83$ mW/cm^2

Table 1 Performance details of devices with Au, Ag, Al and Cu electrodes

Device	R (A/W)[a]	V_{th} (V)[b]	EQE (%)	μ_{dark} (cm^2/V s)[c]	μ_{ill} (cm^2/V s)[c]
Au electrodes	14.5×10^{-2}	−1.5	27.63	42.3×10^{-3}	102.4×10^{-3}
Ag electrodes	3.3×10^{-2}	−2.4	6.25	6.3×10^{-3}	7.2×10^{-3}
Al electrodes	3.7×10^{-2}	−4.1	7.16	4.3×10^{-3}	4.6×10^{-3}
Cu electrodes	1.3×10^{-2}	−4.8	2.49	2.1×10^{-3}	4.4×10^{-3}

[a]The photoresponsitivity (R) of devices obtained at $V_{GS} = -50$ V, $V_{DS} = -50$ V
[b]The threshold voltage (V_{th}) of devices obtained in the dark.
[c]μ_{dark} is the mobility in the dark and μ_{ill} is the mobility under illumination.

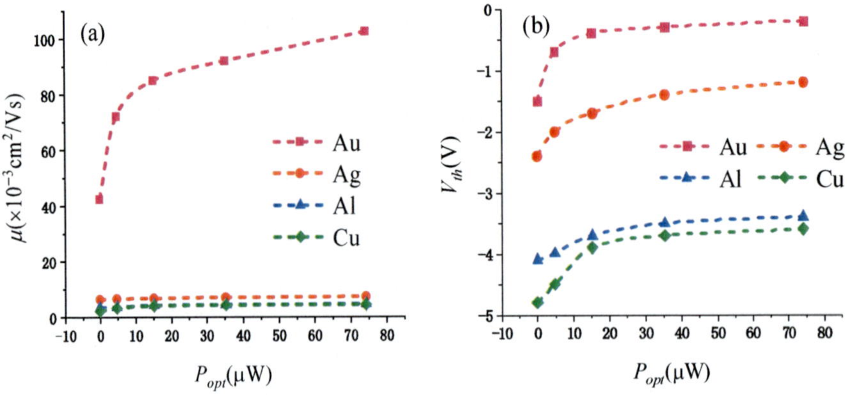

Fig. 3 **a** The carrier mobility (μ) and **b** The threshold voltage (V_{th}) as a function of P_{opt}

Here W and L are the channel width and length, hc/λ is the incident photon energy, I_{DS} is the output current in the saturation region, and C_{ox} is the capacitance per unit area of the insulator, its value is 3.9 nF/cm^2. These results are shown in Table 1 and Fig. 3.

From Table 1, it can be clearly observed that R and EQE of the OPT with gold electrodes presents the highest value among all the devices. While this performance of the OPT with copper electrodes presents the lowest value, which is only about 10% of gold. Meanwhile, the device with gold electrodes show the lowest V_{th} and then it performs best in mobility, which is about 10 to 20 times than other devices.

As shown in Fig. 3, with the increase of light intensity, V_{th} of the four different metal electrodes devices are shifted toward the forward voltage, and the mobility of these four devices are also increasing. This is because as the light intensity increases, a large number of photo-excitons are generated in the organic functional layer. The photo-excitons are dissociated into electrons and holes, and the holes from a current under the action of the source-drain voltage, which strengthens the source-drain current and then increases the carrier mobility of the devices. The accumulation of electrons at the source increases the width of the conductive channel between source and drain, and reduces the potential barrier between the source and the channel.

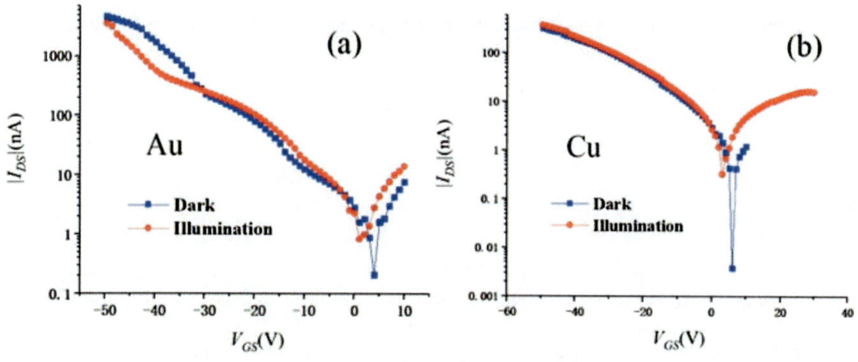

Fig. 4 Transfer characteristics of **a** gold and **b** copper electrodes devices on a logarithmic coordinate axis in dark and under illumination

Therefore, the direct contact resistance between electrodes and organic functional layer is reduced, thereby reducing the threshold voltage.

Through the comparison of the above performance, the gold electrode may be the most suitable for CuPc OPTs. But when we noticed the photo/dark current ratio P, which can be calculated from Eq. 5, the gold electrode lost its advantage.

$$P = \frac{I_{photo}}{I_{dark}} \tag{5}$$

When I_{dark} takes the minimum, P will be the maximum P_{max}. From Fig. 4a, the device with gold electrodes has a minimum current of 0.211 nA in the dark, and the corresponding V_{GS} is 3.98 V. At this voltage, the photocurrent is 12.632 nA. It can be calculated that P_{max} of gold electrodes OPT is 59.9. Using the same method, the device with copper electrodes has a P_{max} of 488.5 from Fig. 4b. This value is much bigger than gold electrodes. This phenomenon may be related to the work function difference between the metal electrode and the organic semiconductor material. The work function of gold is 5.1 eV, and CuPc is 5.3 eV. In contrast, the difference is smaller than copper and CuPc, when the work function of copper is 4.65 eV. Therefore, in the device with copper electrodes, the potential barrier between the organic functional layer and the electrode is higher. This results in a smaller I_{dark} of the device in dark, and P_{max} becomes larger.

4 Conclusions

We confirmed that this CuPc-based OPTs device cannot obtain the optimal photo/dark current ratio and maximum photoresponsitivity by changing the material of the top electrodes. The performance of the device with gold electrodes, such as photoresponsitivity, threshold voltage and mobility, showed better than other device. Conversely, the device with copper electrodes showed better on photo/dark current ratio.

References

1. Shiono F, Abe H, Nagase T, Kobayashi T, Naito H (2019) Optical memory characteristics of solution-processed organic transistors with self-organized organic floating gates for printable multi-level storage devices. Org Electron 67:109–115
2. Smits ECP, Mathijssen SGJ, Hal PAV, Setayesh S, Geuns TCT, Mutsaers KAHA, Cantatore E, Wondergem HJ, Werzer O, Resel R, Kemerink M, Kirchmeyer S, Muzafarov AM, Ponomarenko SA, de Boer B, Blom PWM, de Leeuw DM (2008) Bottom-up organic integrated circuits. Nature 455:956–959
3. Nielsen CB, Turbiez M, McCulloch I (2013) Recent advances in the development of semiconducting DPP-containing polymers for transistor applications. Adv Mater 25(13):1859–1880
4. Takimiya K, Nakano M, Sugino H, Osaka I (2016) Design and elaboration of organic molecules for high field-effect-mobility semiconductors. Synth Met 217:68–78
5. Jeon J, Tee BCK, Murmann B, Bao ZN (2012) Micro-imprinted prism substrate for self-aligned short channel organic transistors on a flexible substrate. Appl Phys Lett 100(4):043301

Simulation Software Development for Charge Transport Characteristics in Organic Semiconductors Based on VB and Fortran Mixed Programming

Changfeng Gu, Yingquan Peng, and Wenli Lv

Abstract In this paper, we report on the mixed programming software of VB and Fortran to determine the charge transport characteristics of organic semiconductors. Exponentially distributed trap and Frenkel effect space charge limited current (SCLC) model was studied respectively. In order to verify the effectiveness of the simulation software, we fabricated and characterized single-hole devices based on organic semiconductor NPB (N,N′-Bis (naphthalen-1-yl)-N,N′-bis (phenyl) benzidine). The simulation current–voltage characteristic curves of the device were consistent with the experimental ones.

Keywords VB · Fortran · Simulation software · Charge transport

1 Introduction

In recent years, organic thin film transistors, organic electroluminescent diodes and other related organic semiconductor materials and devices have made breakthrough development [1]. Charge transport has an important influence on device performance. At present, there are thousands of research work about the development of organic semiconductor materials and devices. There are few reports on simulation software for charge transfer characteristics of organic semiconductor.

Due to programs of Fortran would run faster than any other high-level programming language, Fortran is widely used in engineering calculations for scientific operations. But when it comes to Visual interface design, Fortran has no advantage over other visual programming languages like Visual Basic. As a Windows software development language, VB has strong visual graphics functions and high development efficiency. It uses object-oriented programming ideas to decompose complex design

C. Gu · Y. Peng (✉) · W. Lv
College of Optical and Electronic Technology, China Jiliang University, Hangzhou, People's Republic of China
e-mail: yqpeng@cjlu.edu.cn

Y. Peng and X. Dong (eds.), *Proceedings of 2019 International Conference on Optoelectronics and Measurement*, Lecture Notes in Electrical Engineering 726,
https://doi.org/10.1007/978-981-33-4110-4_14

problems into a collection of objects that can complete independent functions. VB graphics function is used to realize the visualization of Fortran calculation program [3]. Write computation code in Fortran to generate DLL (Dynamic Link Library) files, then use VB language to call DLL files to achieve its computation function. The call between Fortran and VB is basically achieved through the file search call, thus achieving hybrid programming.

Parameters of the charge transport property are mainly the carrier mobility, the density of trap states and their energy distribution. Space charge limited current (SCLC) model was used to determine the mobility of the semiconductor in this paper. The SCLC model is very suitable for thin film devices with a thickness of 100 nm. In organic semiconductors, due to the low mobility, the carrier concentration is usually relatively high, thereby forming a space charge. These space charges generate an electric field which suppress or limit the conduction of current. Therefore, the carrier charge, electric field and current are interrelated and restricted [2]. The SCLC method for measuring mobility mainly uses the relationship between the space charge limiting current and the applied voltage. The Mott-Gurney law states that the current density satisfies the following formula [4] without any trapping effect:

$$J = \frac{9}{8}\mu\varepsilon_r\varepsilon_0\frac{U^2}{d^3}, \tag{1}$$

where μ is the mobility of free carriers, ε_0 is the vacuum dielectric constant, ε_r is the dielectric constant of the material, U is the applied voltage, and d is the distance between the two electrodes or the thickness of the film.

Under ohmic injection conditions, the charge-transport characteristics of an organic semiconductor can be obtained by measuring the current–voltage curves of its single-layer device. This paper develops simulation analysis software for current–voltage characteristics of organic semiconductor single-layer devices to predict the charge-transport characteristics of organic semiconductors.

2 Model and Software Design

2.1 Space Charge Limited Current (SCLC) Model

The mobility is obtained by comparing the fitted curve programmed by the calculation formula with the actual experimental data. According to literature reports, for the case where the density of the trap state in the organic layer is exponentially distributed with its energy and where the ohmic injection is satisfied, the current–voltage characteristic curve of the organic diode satisfies the following formula [5]:

$$I = A\mu q^{1-l}(\frac{l\varepsilon_0\varepsilon_r}{N_t(l+1)})^l(\frac{2l+1}{l+1})^{l+1}\frac{(U-U_{bi})^{l+1}}{d^{2l+1}}, \tag{2}$$

where $A = 4\ \text{mm}^2$ is the effective area of each organic diode; μ is the carrier mobility to be measured, here hole mobility; q is the basic charge; l is the reduced characteristic trap depth; U_{bi} is a built-in potential generated inside the device.

For the case of space charge with Frenkel effect limited current conduction, mobility increases with the square root index of the electric field [6], the relationship between current density and voltage can be expressed as:

$$J = \frac{9}{8}\mu_0\varepsilon_r\varepsilon_0 e^{0.89\beta\sqrt{F}}\frac{F^2}{d} \tag{3}$$

Equation (3) is the mathematical model used in this paper for simulation after multiple equations are integrated and calculated. Among them, β is an exponential factor, and μ_0 is called a zero-field mobility. Using the formula $\mu = \mu_0 e^{\beta\sqrt{F}}$, the value of mobility can be calculated directly from β and μ_0. F is the average electric field and U is the applied voltage. The final measured value during the experimental process is the voltage-current characteristic curve, so the fitted current density is converted into a current, $I = J \times A$.

2.2 *Software Design*

A simulation software for analyzing charge transport characteristics in organic semiconductors was designed utilizing VB and Fortran Mixed Programming. The display interface of the software was shown in Fig. 1. There are two buttons on the main interface of the program: "Exponentially distributed trap SCLC" and "Frenkel effect SCLC" corresponding to the SCLC with exponential distributed trap states and SCLC with Frenkel effect.

Figure 2 shows the interface opened by clicking the button "Exponential Distribution SCLC". For a certain device, anode, cathode and organic semiconductor can be chosen in the interface, and work functions of electrodes and energy level of organic semiconductor should be entered. The input parameters include mobility, thickness, active area, total trap concentration and built-in potential donated by mue, d, N_t and U_{bi} in the interface respectively. A ".txt" file of the measured IV characteristic curve was read by the software.

3 Results and Discussion

In order to analyze the hole transport characteristics of organic semiconductor NPB thin films, a single hole device as shown in Fig. 3 was fabricated. The IV curve of

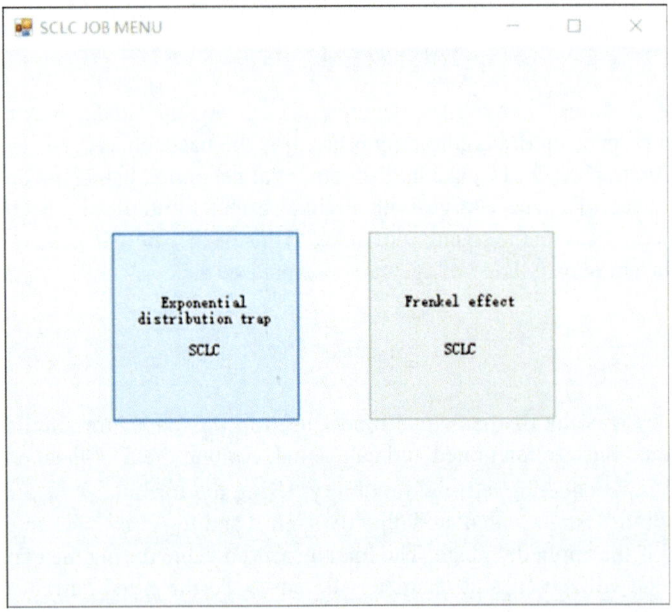

Fig. 1 SCLC software analysis interface

the device was measured, and then the software module 1: exponential distribution trap SCLC was used to analyze the hole transport characteristics. The premise of using the SCLC model is that the metal–organic semiconductor interface must be ohmic contact. The role of MoO_3 is to achieve ohmic contact between the organic semiconductor NPB and the ITO electrode (Fig. 4).

Data fitting was done for 100-nm-thick device and 200-nm-thick device respectively. For the simulation, parameters A, q, ε_0, ε_r, U_{bi}, and d were set as fixed values, and the values of μ, N_t, and l were changed. *Chi-sq* is calculated according to the following formula (4),

$$chi\_sq = \sum_m \left[i_{ex,m}(v_m) - i(v_m)\right]^2 \tag{4}$$

where i_{ex} and i is respectively the measured value of dimensionless and the theoretical value calculated by the Eq. (2). By the simulation, the group (μ, N_t, l) that makes *chi-sq* smallest is found. The minimum value of *chi-sq* reflects the degree of conformity between the device's *IV* characteristics and model (1). The smaller the *chi-sq*, the better the degree of conformity. The final experimental data curve is close to the calculated fitting curve, and the mobility with less error is obtained. Here, some parameters are generally set as fixed values, $A = 4$ mm^2; $\varepsilon_r = 3$; $U_{bi} = 0.9$ V.

The red curve in the Fig. 5 is obtained by the software simulation, while the black point is the experiment data. It can be seen from the fitting results of organic diodes

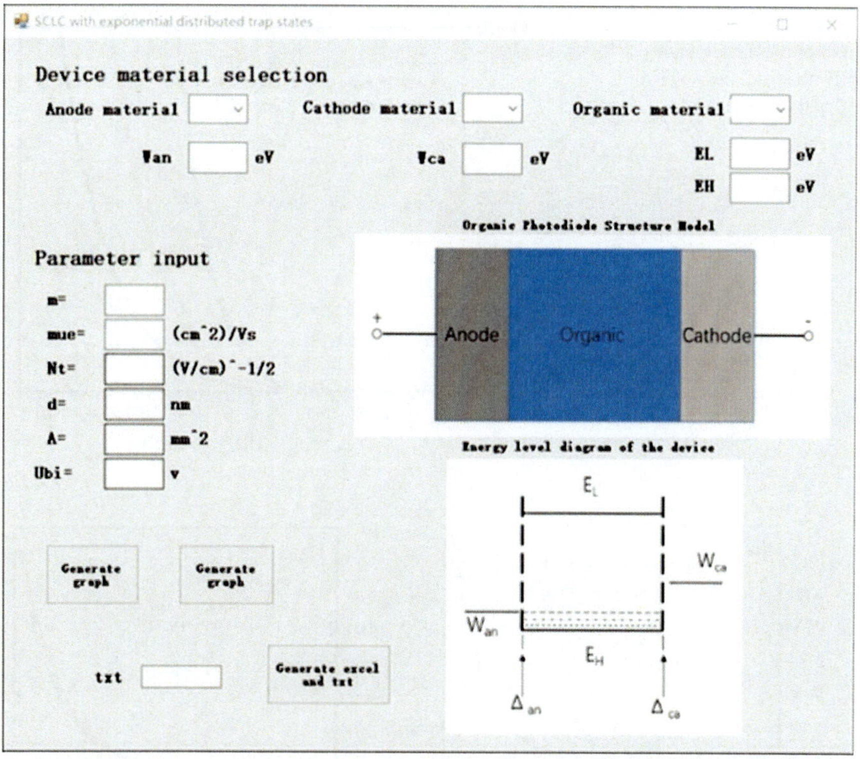

Fig. 2 SCLC with exponential distributed trap states

Fig. 3 Organic diode structure: ITO/MoO_3 (3 nm)/NPB (50 nm; 100 nm; 150 nm; 200 nm)/BCP (10 nm)/Al

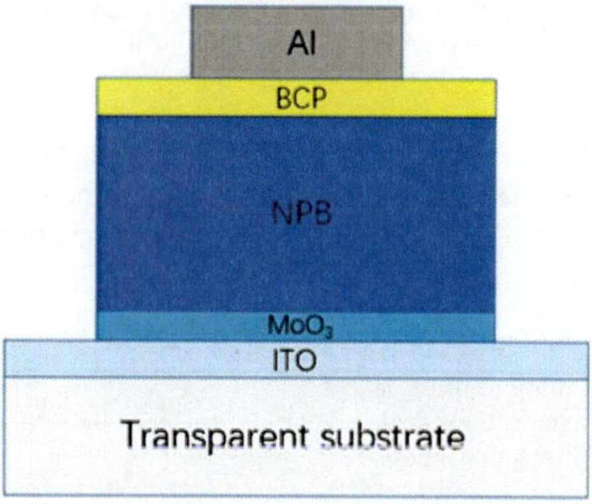

that the experimental data are basically consistent with the calculated curve, and the hole mobility is between 7×10^{-5} cm^2/Vs and 9×10^{-5} cm^2/Vs, which is not much different from the reported 1.1×10^{-5} cm^2/Vs to 3.5×10^{-4} cm^2/Vs [7]. This indicates that formula (2) and formula (2)–(3) correctly reflect the voltage and current

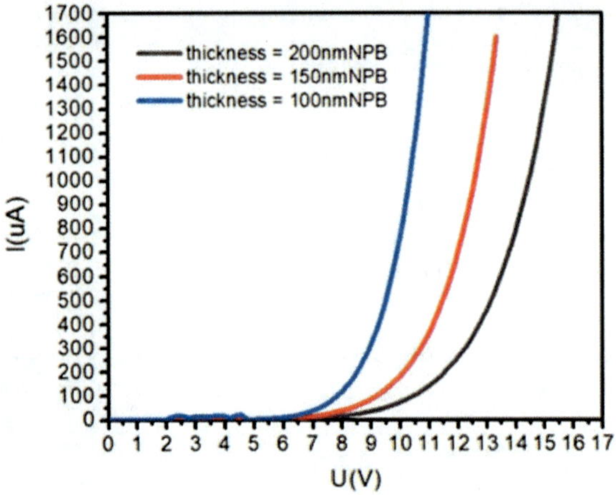

Fig. 4 IV characteristic curve of the device with NPB film thickness of 100, 150 and 200 nm

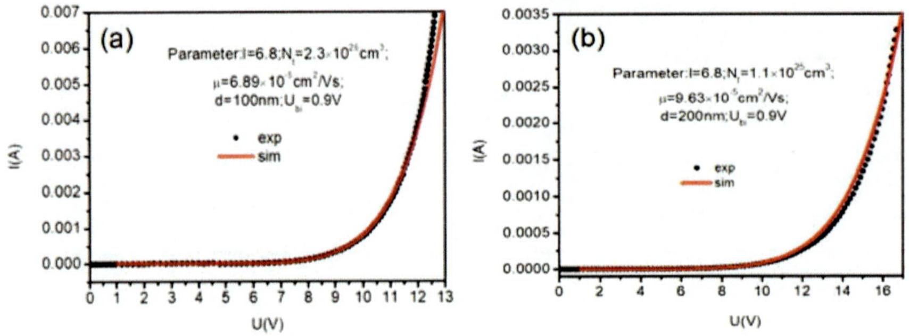

Fig. 5 Fitting curves of two thickness organic diodes. **a** fitting curves of 100 nm thickness devices. **b** fitting curve of 200 nm thickness device

characteristic curve of organic diodes, and the mobility of organic materials can be obtained by this fitting method. As shown in Fig. 5, for the device with NPB layer thickness of 100 and 200 nm, the simulation results of μ, Nt, and l was summarized in Table 1.

Table 1 Fitting data of organic diode devices with different thickness

Thickness (nm)	I	N_t (cm^3)	μ (cm^2/Vs)
100	6.8	2.3×10^{25}	6.89×10^{-5}
200	6.8	1.1×10^{25}	9.63×10^{-5}

4 Conclusion

A simulation software based on the mixed programming of VB and Fortran was studied to determine the charge transport characteristics of organic semiconductors. Single-hole diodes based on NPB were fabricated and characterized. The simulation current–voltage characteristic curves of the device were consistent with the experimental curves, which indicates that the SCLC model can accurately characterize the charge transport characteristics of the devices.

References

1. Chen ZY, Ye TL, Ma DG (2009) Measurement methods for charge carrier mobility in organic semiconductors. Progressin Chem 21(5):940–947
2. Xv KF, Zhang DG (2014) Measurement of hole mobility of organic material based on space charge limited currents. Mech Electr Eng Technol 43(9):44–48
3. Zhao RC, Wang YY, Li YL (2011) Mixture programming of Fortran and VB on working process calculation for diesel engine. SHIP BOAT 22(4):70–73
4. Murgatroyd PN (1970) Theory of space-charge-limited current enhanced by Frenkel effect. J Phys D Appl Phys 3(2):151
5. Burrows PE, Forrest SR (1994) Electroluminescence from trap-limited current transport in vacuum deposited organic light emitting devices. Appl Phys Lett 64(17):2285–2287
6. Yeargan JR, Taylor HL (1968) The Poole-Frenkel effect with compensation present. J Appl Phys 39(12):5600–5604
7. Ishihara S, Hase H, Okachi T et al (2012) Simulation of impedance spectra of double-layer organic light-emitting diodes for the determination of hole drift mobility of NPB/Alq3 diodes by means of impedance spectroscopy. Physica Status Solidi 9(12):2561–2564

Review of Magnetic Fluid Fiber Magnetic Field Sensing Technology Based on Interference Mechanism

Yuchan Liu, Yueming Liu, and Cheng Xu

Abstract Magnetic fluid is a new type of optical functional material, which has excellent magneto-optical properties in magnetic field media. On this basis, many scholars have combined it with fiber-optic sensing technology to make a variety of fiber-optic magnetic field sensors based on interference mechanism. The technical performance of magnetic fluid fiber magnetic field sensors based on interference mechanism is analyzed, and the research status and progress of magnetic fluid-based fiber-optic magnetic field sensors are summarized and prospected in this article.

Keywords Magnetic fluid · Fiber optic sensor · Magnetic field measurement · Review

1 Introduction

With the rapid development of the sensing technology, the magnetic field sensors have been widely used in aerospace, medicine, submarine environment detection, geological exploration, and power systems, et al. [1–3], and play an important role in these fields. Traditional magnetic sensors are usually electrical signal types, such as Hall magnetic sensors, magnetic-resistance sensors, and fluxgate sensors [4–6] which usually use metal wires to provide current excitation. The excitation current easily interferes with the magnetic field distribution which limits the improvement of detection accuracy. In addition, traditional magnetic sensors are usually large in size, cannot be miniaturized and have relatively large power consumption. The optical fiber magnetic field sensors inherit the advantages of the optical fiber sensor including small dimension, corrosion resistance, high anti-electromagnetic interference capability, convenient distributed multi-point measurement, and all-optical transmission, which have now become one of the main research directions of magnetic field sensing

Y. Liu · Y. Liu (✉) · C. Xu
College of Optics and Electronic Technology, China Jiliang University, Hangzhou 310018, China
e-mail: liuym@cjlu.edu.cn

Y. Peng and X. Dong (eds.), *Proceedings of 2019 International Conference on Optoelectronics and Measurement*, Lecture Notes in Electrical Engineering 726,
https://doi.org/10.1007/978-981-33-4110-4_15

technology. According to different sensing principles, fiber-optic magnetic field sensors mainly have the following research directions: (1) A magnetic field sensor is achieved by coating a thin film of magnetostrictive material on the surface of the optical fiber [7, 8], (2) A magnetic field sensor is achieved by using the magnetic rotation effect of magneto-optical crystal, but it is greatly affected by the environment and polarization [9], (3) A magnetic field sensor is achieved by the combination of a cantilever beam and a fiber grating, but it is easily interfered by external media [10], (4) A magnetic field sensor is achieved by utilizing the adjustable refractivity of magnetic fluid. The magnetic field sensor based on magnetic fluid has become a research focus in recent years due to the magnetic fluid has no mechanical wear and no moving parts [11–14].

Magnetic fluid, also known as a magnetic liquid, ferro fluid or magnetic fluid, is a colloidal solution formed by magnetic nanoparticles uniformly dispersed in the base carrier fluid under the encapsulation effect of the surfactant [15]. Magnetic fluid not only has the fluidity of liquid but also the magnetic properties of solid magnetic substances, as well as rich optical properties including thermal lens effect, magneto-induced birefringence effect, and adjustable refractive index characteristics [16, 17]. Based on the adjustable refractivity of magnetic fluid, the magnetic field sensor has been used widely. The refractivity of the magnetic fluid varies with the changes of the magnetic field, showing different trends with the direction of a magnetic field. In addition, the refractive index of a magnetic fluid is also affected by temperature, which will expand the application range of magnetic field sensors based on magnetic fluid in the measurement field.

The development and the current state of magnetic fluid fiber magnetic field sensing technology based interference mechanism are summarized in this article. According to different sensing mechanisms, the sensors are divided into four categories: fiber magnetic field sensors based interference mechanisms which include Fabry–Perot interference [11–13], Sagnac interference [18–23], Mach–Zehnder interference [24], and mode interference mechanism [25–28]. The advantages and disadvantages of sensor mechanism, sensor structure and sensor performance are compared. Finally, the development of the magnetic field sensors based magnetic fluid is summarized and prospected in this article.

2 Optical Fiber Magnetic Field Sensor Based on an Interference Mechanism

2.1 *Based on the F-P Interference Mechanism*

As a multi-beam interferometer, F-P interferometer is a reflection-type interferometer which can be combined with a magnetic fluid to realize a magnetic sensor. The interference signal can be expressed as $I^{(r)} = I_1 + I_2 + 2\sqrt{I_1 I_2} \cos\delta$. Here I_1 and I_2 are the intensity of light reflected from the first and second optical planes, respectively, and δ is the phase difference in the F-P cavity which can be expressed as $\delta = 4\pi nL/\lambda$. Here n is the refractive index of the magnetic fluid in the F-P cavity, and L is the cavity length of the F-P cavity and λ is the wavelength of the output light. Therefore, the output signal of the F-P cavity varies with the refractive index of the magnetic fluid under the action of the magnetic field, and the magnitude of the magnetic field can be measured by measuring the change of the wavelength of output light. In 2012, Zhao et al. proposed a magnetic field sensor [11] to fill the magnetic fluid using the capillarity of HC-PCF. The HC-PCF filled with magnetic fluid was used as the F-P interference cavity. The measured magnetic field sensitivity of the sensor in the range of 50–150 Oe was 33 pm/Oe. Different F-P cavity lengths had different output spectrums, so the sensor has better multiplexing capability.

In 2014, Lv RQ et al. proposed a magnetic field sensor based on a magnetic fluid-filled fiber F-P cavity [12]. The end faces of two single-mode optical fibers constituted an F-P cavity was filled with magnetic fluid. The sensor had a magnetic field sensitivity of 43.1 pm/Oe in the range of 0–400 Oe. On this basis, Wang Dan of Northeastern University ruled FBG on one of two single-mode fibers, which solved the problem of cross-sensitization of magnetic field and temperature and improved the accuracy of magnetic field measurement [13]. The structure is shown in Fig. 1. The magnetic field sensitivity of 34 pm/Oe and the temperature sensitivity of 0.013 nm/°C were obtained. The sensor can be used in the environment where the magnetic field and temperature change at the same time.

In 2018, Fuquan Shi et al. proposed an F-P interference magnetic field sensor [14]. The structure is easy to manufacture, as shown in Fig. 2. The sensing mechanism is the characteristics of magnetic nanoparticles arranged along the direction of the magnetic field. When there is a magnetic field, the coreless fiber shifts slightly with

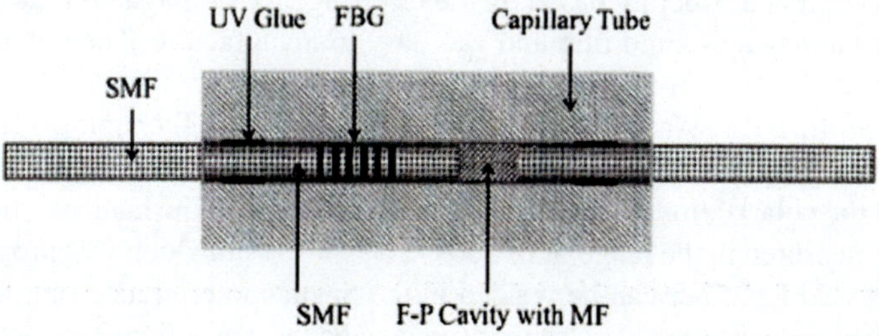

Fig. 1 Structure with temperature compensation. Reprinted with permission from Ref. [13]. Copyright 2014 cnki.net

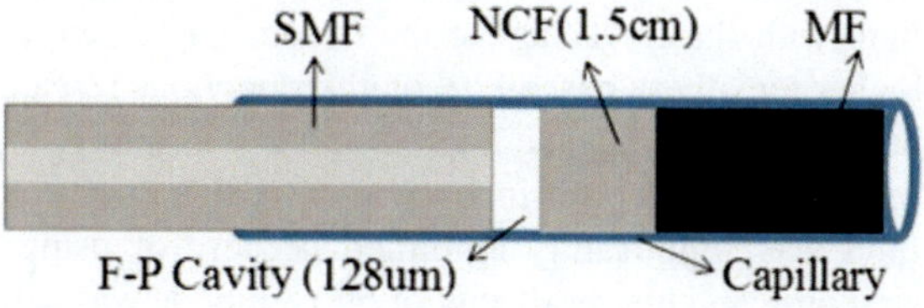

Fig. 2 Structure with temperature compensation. Reprinted with permission from Ref. [14]. Copyright 2018 Optical Fiber Technology

the change of the volume of the magnetic fluid. At the same time, the output optical signal will also be changed with the change of the air cavity length, thus measurement of the magnetic field can be achieved. When the direction of magnetic field was parallel to the capillary placement direction, the measured magnetic field sensitivity was 0.02347 nm/Oe. When the magnetic field direction was perpendicular to the placement direction of the capillary, the magnetic field sensitivity was 0.0325 nm/Oe.

2.2 Based on the Sagnac Interference Mechanism

The Sagnac interference structure is composed of a 3 dB single-mode coupler whose two output ends are welded together to form a ring structure. To improve the sensitivity of the sensor, a polarization-maintaining fiber is generally welded in the ring. The Sagnac interference structure can be easily combined with various other fiber-optic sensing structures to achieve high-sensitivity magnetic field measurements. Sensing mechanism of fiber-optic magnetic field sensors based on the Sagnac interference is the magnetically induced birefringence characteristics of the magnetic fluid. The reflection function of Sagnac interference ring can be expressed as $R\lambda = 1 - [\sin\theta \cdot \cos 2\delta]^2$·. Here δ is the phase difference and can be expressed as $\delta = \pi BL/\lambda$. Here L is the equivalent length and λ is the wavelength, and B is birefringence which includes birefringence B_0 of polarization-maintaining fiber and birefringence B_m of magnetic fluid. B_0 is determined by the polarization-maintaining fiber and is a constant. B_m will be changed with the magnetic field and the signal of the output light will also be changed. The magnetic field can be measured by measuring the displacement of the output spectrum. In 2011, Zu et al. proposed a Sagnac interference structure based on the birefringence of magneto fluid [18]. The sensor used a magnetic fluid film and was easy to manufacture. The magnetic field sensitivity of the 60 μm-thick magnetic fluid film in the range of 0–180 Oe was 16.7 pm/Oe. In 2014, Wang Jing of WuHan University of Technology proposed a fiber-optic magnetic field sensor based on the Sagnac interference ring [19] in which the clad of the polarization-maintaining fiber was corroded. The magnetic field sensitivity was measured in the range of 0–200 Oe was 11.31 pm/Oe. It was proposed that an unprocessed PM fiber can be welded in the Sagnac interference ring to achieve temperature compensation. The sensor can also be used as a liquid refractive index sensor. Organic glass material was used to package the sensing probe and magnetic fluid in the experiment. The volume of the sensor after packaging was larger than that of ordinary sensors because of the ring structure.

In 2014, Zu et al. proposed that combining the birefringence effect of the magnetic fluid with the Loyt-Sagnac interferometer to achieve a magnetic field sensor had higher sensitivity based [20] on their previous research [21, 22]. As shown in Fig. 3 below, the Loyt-Sagnac interferometer was a Sagnac interferometer containing two segments of high birefringence fiber (HBF). The birefringence effect of the magnetic fluid was significantly amplified because of using the Loyt-Sagnac interference structure, and the sensitivity of 592.8 pm/Oe was achieved.

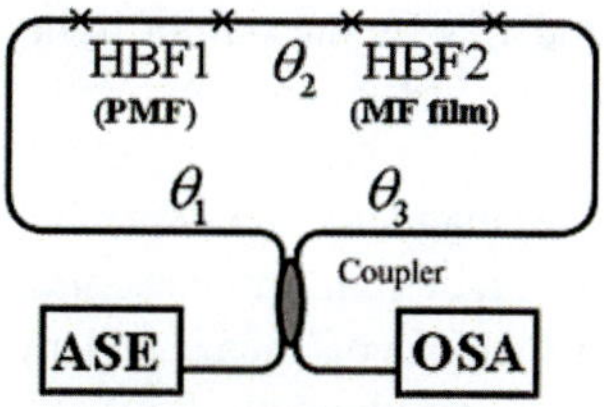

Fig. 3 Loyt-Sagnac interference structure. Reprinted with permission from Ref. [20]. Copyright 2014 Sens.Actuators B: Chem

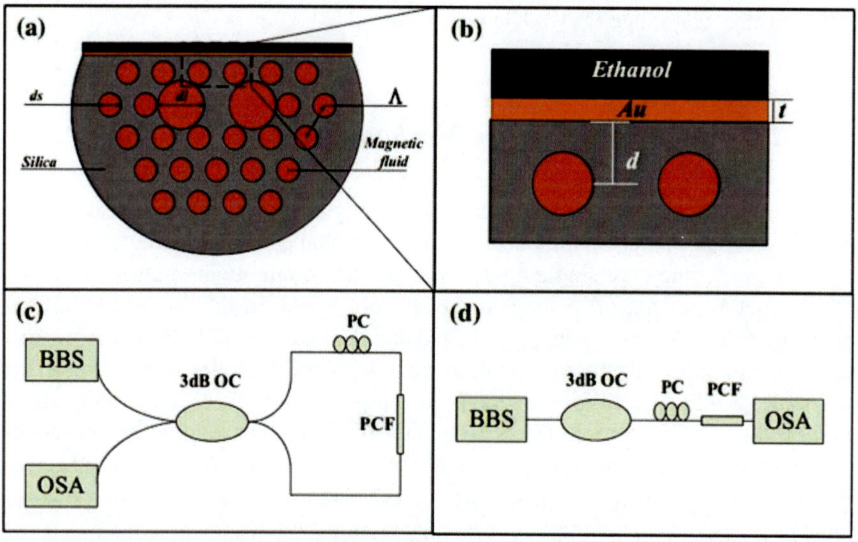

Fig. 4 **a**, **b** are cross-sections of photonic crystal fibers; **c** is a system diagram of Sagnac interference; **d** is a schematic diagram of the surface plasmon resonance effect. Reprinted with permission from Ref. [23]. Copyright 2018 Optical and Quantum Electronics

In 2018, Hai Liu et al. proposed a sensor based on Sagnac interference structure and D-type photonic crystal fiber [23], as shown in Fig. 4. The air holes of photonic crystal fiber were filled with magnetic fluid and the D-shaped area was coated with a gold film. The magnetic field sensitivity was and the sensitivity was effectively improved because of the combination of Sagnac interference structure and surface plasmon resonance effect. In the experiment, the magnetic field sensitivity of 0.483 nm/Oe and the temperature sensitivity of 0.1 nm/°C were achieved.

2.3 Based on the Mach–Zehnder Interference Mechanism

The Mach–Zehnder interferometer is also a different interference-based structure from the F-P interferometer, which is a transmissive sensor that the transmission signal can be expressed as $\mathrm{I} = I_{out1} + I_{out2} + 2\sqrt{I_{out1} I_{out2}} \cos(\delta + \delta_0)$. Here I_{out1}

and I_{out2} is the output light intensities of the two cores, and δ_0 is the initial phase difference. δ can be expressed as $\delta = 2\pi L\Delta n/\lambda$. Here L is the length of micro-cavity and Δn is the refractive index difference, and λ represents the output light wavelength. The magnetic measurement of the Mach–Zehnder interference sensors based on magnetic fluid can be achieved using the change of refractivity of the magnetic fluid. In 2015, Li of ShenZhen University proposed a Mach–Zehnder interference structure included magnetic fluid and the dual-core fiber in which a microcavity was made [24]. The magnetic field sensitivity of −37.1 nm/Oe was achieved, but the magnetic field measurement range of 0–80Oe was achieved.

2.4 *Mode-Based Interference Mechanism*

Previous studies are various about fiber-optic magnetic field sensors based on the mode interference mechanism. Such as sensors with tapered shape structure, sensors of single mode-fine core-single mode structure and single mode-multimode-single mode structure, et al., all belong to the sensors based on the mode interference mechanism. The sensor is generally melted and tapered or made a core-offset structure, and the core fundamental mode excites high-order modes at a fusion plane and mode coupling occurs at the second fusion plane, so which is essentially also a sensor of Mach–Zehnder interference mechanism. When the two interference modes meet certain phase conditions which is $2\pi\left[n_{eff}^{co}(\lambda) - n_{eff}^{cl,j}(\lambda, n_{ext})\right](L/\lambda_D) = (2k + 1)\pi$, there will be a transmission valley which is affected by the refractive index of the external environment. When the external environment is the magnetic, the refractive index of magnetic fluid will change. So the magnitude of the magnetic field can be obtained by measuring the drift of the interference trough. Here n_{eff}^{co} and $n_{eff}^{cl,j}$ is the effective refractive index of the fundamental mode and the j-th order cladding mode, respectively, and n_{ext} is the effective index of the environment, and L is the length of the sensing area, and λ_D is the wavelength of the interference valley, and k is integer. In 2014, Layeghi and others proposed a magnetic field sensor with a tapered structure [25]. The tapered fiber structure was coated with a layer of magnetic fluid and sealed in the capillary tube, which can be made small. The magnetic field sensitivity of −71.7 pm/Oe was achieved. Shengli Pu et al. proposed a sensor with a tapered joint [26], as shown in Fig. 5. The magnetic field sensitivity of 32.53 pm/Oe was achieved ranging from 0 to 160 Oe.

In 2016, Liang of Zhejiang University achieve magnetic field measurement combined cladding-etched thin-core fiber and magnetic fluid [27]. The sensor can achieve dual parameter measurements of the magnetic field and temperature, and the magnetic field sensitivities of 128 pm/Oe and the temperature sensitivities of 12.23 pm/°C were achieved. The structure is shown in Fig. 6.

In 2017, Yue designed a fiber-optic magnetic field sensor based on peanut cone and multimode fiber structure [28]. The two ends of the single-mode fiber engraved long-period fiber grating and the multimode fiber were melt-balled to make peanut

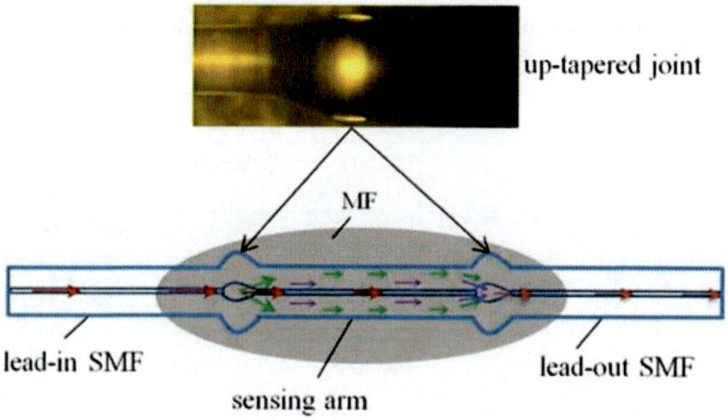

Fig. 5 Sensor with tapered joint. Reprinted with permission from Ref. [26]. Copyright 2014 IEEE Photonics journal

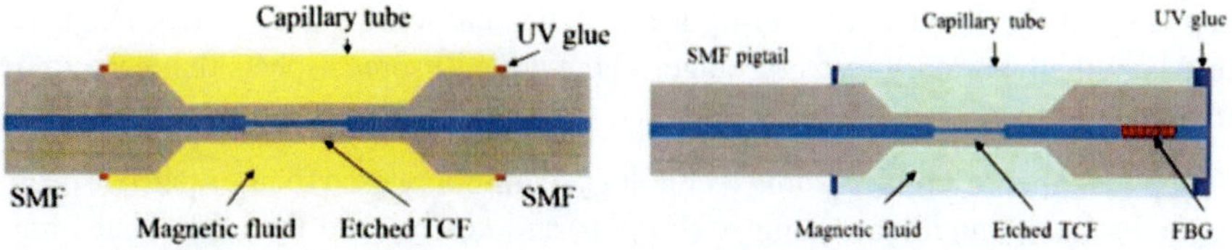

Fig. 6 Schematic diagram of TCFMI structure of cladding corrosion by magnetic fluid coating (right: Schematic diagram of TCFMI structure of cascaded FBG). Reprinted with permission from Ref. [27]. Copyright 2016 cnki.net

cone structures. In the experiment, the magnetic field sensitivity of 47.8 pm/Oe and the temperature sensitivity of 0.06 nm/°C were obtained.

3 Summary

The magnetic field sensitivity, measurement range, and magnetic field measurement mechanism of fiber-optic magnetic field sensors based different interference mechanisms are summarized, as shown in Table 1. The measurement mechanism of magnetic field sensors is generally achieved by measuring the change of wavelength. For the packaging of sensors, which is packaged with an organic glass when the sensing probe is a ring structure, for others sensors, with a capillary for packaging. For the filling of magnetic fluid, previous studies achieves filling through capillary action or the pressure of syringe.

Table 1 Comparison of sensing performance of fiber-optic magnetic field sensors based on different interference mechanisms

Interference structure	Sensitivity(pm/Oe)	Measuring range (Oe)	Measurement mechanism	Packaging method
F-P interference	33–43.1	0–400	Wavelength shift	Quartz glass capillary
Sagnac interference	16.7–580	0–200	Wavelength shift	Plexiglass material
Mach Zander interference	37,100	0–8	Wavelength shift	Quartz glass capillary
Mode interference	47.8–128	0–440	Wavelength shift	Quartz glass capillary

4 Summary and Prospect

At present, the technology of fiber-optic magnetic field sensing based on the magnetic fluid is in the laboratory research stage. There are still some aspects that need to be improved:

1. Optical fiber Micromachining technology, commonly used focuses particle beam or Fs Laser micromachining is complex and costly. And the traditional semi-conductor planar micromachining technology is not suitable for the three-dimensional micromachining of optical fibers, it is necessary to develop technological methods suitable for optical fiber three-dimensional micromachining.
2. The effective filling of the magnetic fluid is a difficult point of the micro-structure fiber sensor technology, which determines the performance of the sensor to a large extent, so it is necessary to develop more effective and low-cost magnetic fluid filling methods.
3. The packaging technology of fiber magnetic field sensors based on magnetic fluid needs to be strengthened and improved, which determines the performance and lifetime of the sensor, as well as the level of sensor development costs.
4. The magnetic fluid materials which have high viscosity, slow response time and are not easy to fill and package is commonly used. The performance of magnetic fluid materials is a key element of fiber-optic magnetic field sensors. Therefore, new magnetic fluid materials which have faster dynamic response time, proper viscosity, and easy filling and packaging need to be developed.

Consider the temperature characteristics of magnetic fluids, fiber-optic magnetic field sensors based on magnetic fluids are sensitive to temperature. To solve the problem of temperature and magnetic field cross-sensitivity, many research reports have adopted some methods including using fiber grating structure and injecting temperature-sensitive liquid to achieve the dual parameter measurement of the magnetic field and temperature. Temperature-sensitive packaging material can also be used to improve temperature sensitivity. Therefore, magnetic field sensor based on magnetic fluid should not only be limited to measure the magnetic field, but its function can also be extended to the field of temperature measurement to achieve high utilization of the sensor.

Magnetic fluid not only has an adjustable magneto-refractive index and temperature characteristics but also has the magnetic characteristics of solid magnetic materials. The nanoparticles are arranging along the direction of the magnetic field, so the direction of the magnetic field has influence on the magnitude of the magnetic field. At present, the research on magnetic field measurement mostly focuses on the measurement of the magnitude of the magnetic field, while the research on the measurement of the magnetic field direction is less. Consider the difference between the actual environment and laboratory measurement conditions, the direction of the magnetic field must be considered in actual measurement, which is the direction of the magnetic field is known before measuring in the laboratory but the magnitude and direction of the magnetic field are unknown during the actual measurement. Although the study of the direction of the magnetic field is less in previous sensors, the sensors have a certain sensitivity to the direction of the magnetic field because of the magnetic properties of the magnetic fluid. It is hoped that the sensitivity of magnetic field directional of the sensor can be used as an important index to measure the sensing performance to expand the application range of the sensor in the field of magnetic field measurement in future research.

References

1. Xiao XY, Zhou WS (1999) Domestic progress in the biological effects of magnetic fields. Chin J Physiotherapy 3(1):6–9
2. Zhou Z (2006) Effect of magnetic fluid accumulation on the targeting of magnetic drugs. MA. Sc. Theses, Chinese Academy of Sciences, Institute of Electrical Engineering
3. Gitter K, Odenbach S (2011) Quantitative targeting map based on experimental investigations for a branched tube model in magnetic drug targeting. J Magnetism Magnetic Mater 323:3038–3042
4. Ripka P (2003) Advances in fluxgate sensors. Sens Actuators A 106:8–14
5. Hauser H, Hochretter J, Stangl G et al (2013) Anisotropic magnetoresistance effect field sensors. J Magnetism Magnetic Mater 215(216):788–791
6. Paun MA, Sallese JM, Kayal M (2012) Offset and drift analysis of the hall effect sensor. Digest J Nanomater Biostructures 7(3):883–891
7. Yang M, Dai J, Zhou C et al (2009) Optical fiber magnetic field sensors with TbDyFe magnetostrictive thin films as sensing materials. Opt Express 17(23):20777–20782
8. Liu JY, Si YM, Li ZZ et al (2004) Fiber magnetic field sensor using giant magnetostrictive film. Photoelectric Technol Appl 25(03):238–241
9. Zhu J (2013) Research on magnetic field measurement of optical fiber-based on magneto-optical crystal. MA. Sc. theses, Zhejiang University
10. Liu J (2006) MEMS cantilever magnetic field sensor and resonant cantilever beam electromagnetic drive technology. MA. Sc. theses, Shanghai Institute of Microsystem and Information Technology, Chinese Academy of Sciences
11. Zhao Y, Lv R, Ying Y, Wang Q (2012) Hollow-core photonic crystal fiber Fabry-Pérot sensor for magnetic field measurement based on magnetic fluid. Opt Laser Technol 44(4):899–902
12. Lv RQ, Zhao Y, Wang D et al (2014) Magnetic fluid-filled optical fiber Fabry-Pérot sensor for magnetic field measurement. IEEE Photonics Technol Lett 26(3):217–219

13. Wang D (2014) Research on novel fiber optic Fabry-Perot magnetic field sensor based on magnetic fluid. MA. Sc. theses, Northeastern University
14. Shi F, Yan L, Che J et al (2018) Optical fiber F-P magnetic field sensor based on the magnetostrictive effect of magnetic fluid. Opt Fiber Technol 43:35–40
15. Li DC (2003) Magnetic liquid sealing theory and application. Science Press, Beijing, pp 109–152
16. Zhao Y, Dong JL, Li X (2009) Optical properties of magnetic fluids and their applications in the field of photoelectric information sensing. Photoelectric Eng 36(07):126–131
17. Pu S, Chen X, Liao W et al (2004) Laser self-induced thermo-optical effects in a magnetic fluid. J Appl Phys 96(10):5930–5932
18. Zu P, Xiang WH, Bai YB et al (2011) A novel magnetic liquid-based sagnac magnetic field sensor. J Opt 31(08):57–61
19. Wang J (2014) Fiber-optic magnetic field sensor based on magnetic fluid and high birefringence ring mirror. MA. Sc. theses, Wuhan University of Technology
20. Zu P, Chan CC, Koh GW et al (2014) Enhancement of the sensitivity of magneto-optical fiber sensor by magnifying the birefringence of the magnetic fluid film with Loyt-Sagnac interferometer. Sens Actuators B: Chem 191:19–23
21. Zu P, Chan CC, Jin YX et al (2011a) A temperature-insensitive twist sensor by using low-birefringence photonic crystal-fiber-based Sagnac interferometer. IEEE Photonics Technol Lett 23:920–922
22. Zu P, Chan CC, Jin YX et al (2011b) Fabrication of a temperature-insensitive transverse mechanical load sensor by using a photonic crystal fiber-based Sagnac loop. Meas Sci Technol 22(2):0957–1233
23. Liu H, Li H, Wang Q et al (2018) Simultaneous measurement of temperature and magnetic field based on surface plasmon resonance and Sagnac interference in a D-shaped photonic crystal fiber. Opt Quantum Electron 50:392
24. Li ZY (2015) Mach-Zehnder interferometer based on twin-core fiber and sensing applications. MA. Sc. theses, Shenzhen University
25. Layeghi A, Latifi H, Frazer O (2014) Magnetic field sensor based on nonadiabatic tapered optical fiber with magnetic fluid. IEEE Photon Technol Lett 26(9):1904–1907
26. Pu S, Dong S (2014) Magnetic field sensing based on magnetic-fluid-clad fiber-Optic structure with up-tapered joints. IEEE Photonics J 6(4):5300206
27. Zhang L (2016) Magnetic fluid magnetic field sensing based on magnetic fluid. MA. Sc. theses, Zhejiang University
28. Zhao Y (2017) Research on interferometric optical fiber sensor based on peanut cone structure. MA. Sc. theses, Tianjin University of Technology

Controllable Optical Characteristics of TiN Films Prepared by Magnetron Sputtering

Kai Xu, Yan-Long Meng, Si-Meng Liu, and Jun Tan

Abstract In this paper, TiN thin films prepared based on magnetron sputtering are studied. The characteristics of TiN thin films fabricated by non-reactive direct current sputtering are analyzed by using X-ray photoelectron spectroscopy and ellipsometry polarization spectroscopy. By analysis, it is found that there are many N defects in the TiN thin film which results in a low refractive index in the visible range. After changing the manufacture process by introducing N doping, refractive index of the TiN film can be enhanced. The research in this article is helpful to understand the role of N_2 in the deposition process of TiN films, and to provide a feasible solution for the regulation of TiN film properties.

Keywords TiN · Plasmon material · Magnetron sputtering

1 Introduction

As a transition metal nitride material, TiN has been widely used in machine protection and coating protection due to its excellent physical properties, such as high hardness, friction resistance, etc. [1]. Not only that, but the structurally complete TiN material also has a higher electrical conductivity, thus replacing gold in the field of microelectronics, such as diffusion films, gates in field effect transistors, metal connection points in ultra-large-scale integrated circuits, etc. [2–5]. With the deepening of surface plasmon research, TiN has also attracted much attention because

K. Xu · Y.-L. Meng (✉) · S.-M. Liu · J. Tan
College of Optical and Electronic Technology, China Jiliang University, Hangzhou, Zhejiang 310018, P. R. China
e-mail: myl@cjlu.edu.cn

Y.-L. Meng
Key Laboratory for Organic Electronics and Information Displays & Institute of Advanced Materials, Nanjing University of Posts and Telecommunications, Nanjing 210023, P. R. China

Y. Peng and X. Dong (eds.), *Proceedings of 2019 International Conference on Optoelectronics and Measurement*, Lecture Notes in Electrical Engineering 726,
https://doi.org/10.1007/978-981-33-4110-4_16

of its plasma characteristics in the visible light band, and has become a hot topic in recent years to replace traditional noble metal plasma materials [6–8].

So far, various TiN preparation technologies have been developed, such as magnetron sputtering (MS), cathode arc ion plating (CVA), chemical vapor deposition (CVD), atomic layer deposition (ALD), pulsed laser Sedimentation and other technologies [7, 9–11]. However, the characteristics of TiN films are greatly affected by the manufacturing process. Different process conditions in the same technology will also affect the characteristics of TiN films to some extent [12, 13]. Magnetron sputtering has attracted widespread attention as one of the easiest techniques for preparing TiN films. TiN thin films are prepared based on magnetron sputtering. Reactive deposition technology is often used to obtain TiN thin films by sputtering Ti targets in an N_2 atmosphere. However, no research has been reported on the preparation of TiN thin film materials by direct current (DC) sputtering TiN targets. In this paper, TiN films of different thicknesses were deposited on a silicon substrate by DC magnetron sputtering TiN technology. The effects of changing the environmental conditions of the sputtering chamber on the optical properties of TiN and the changes in optical properties and surface morphology of TiN with different thicknesses were studied. The related properties of the thin films obtained by direct sputtering of TiN targets are studied in depth, which provides a favorable reference for DC sputtering TiN technology.

2 Experiments

The substrate used for the preparation of the TiN film was a single-sided polished single crystal silicon wafer. Before preparation, the single crystal silicon wafer was cut into 2 cm × 2.5 cm pieces. After that, the substrate was ultrasonically cleaned in acetone, ethanol, and deionized water for 10 min in subsequence. And then the substrates were dried in an oven, and placed in a sputtering chamber. The sputtering target was a TiN target. When the vacuum in the reaction chamber reaches 3×10^{-4} Pa, we filled the reaction chamber with argon or a mixture of argon and nitrogen (flow ratio 3: 7) required for sputtering, the gas flow is 20 sccm, and the sputtering pressure is 0.86 Pa, with a sputtering power of 80 W. The substrates kept rotating during the sputtering process to ensure uniformity of deposited films. The surface morphology of prepared samples was tested by scanning electron microscopy. X-ray photoelectron spectroscopy (XPS) was used to analyze the elements contained in the materials, and ellipsometry was used to test and analyze the optical characteristics of the materials.

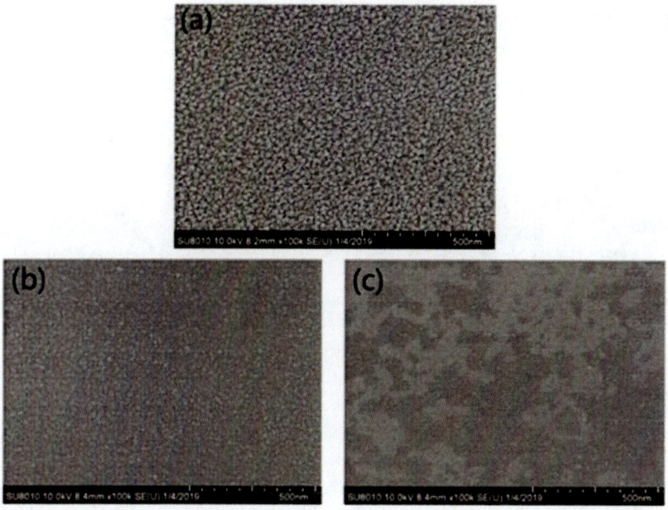

Fig. 1 SEM images of TiN thin films with thicknesses of 100 nm **a** prepared in N_2: Ar = 0: 1 atmosphere and 35 nm **b** 20 nm **c** prepared in N_2: Ar = 7: 3 atmosphere

3 Results and Discussion

Figure 1 shows the surface morphology of TiN under different deposition conditions obtained by SEM testing. It can be seen from the SEM image that even without nitrogen doping, the sample surface still has dense TiN grains, and the grain size of the surface grains is larger than that of a 35 nm-thick TiN film prepared by doping N_2. For the TiN film prepared by doping 70% N_2, the 20 nm-thick TiN film shows obvious discontinuity. In order to further analyze the composition of the TiN thin film not doped with N2, the XPS is used to analyze the Ti, N, C, and O contained in the TiN thin film. The ionization energy spectra of various elements obtained by the test are shown in Fig. 2.

By testing Ti, N, C, and O elements in the thin film sample, it is not difficult to find that the characteristic energy peaks of the C and O elements in the sample are strong, which indicates that C and O adsorption exists on the sample surface, and the presence of the O element may be due to partial oxidation of the film surface. Considering that the standard peak position of the equipment testing C 1s used for the sample test is 284.6 eV, it is known that the sample test has a shift of 0.14 eV by correcting the C 1s peak position. The characteristic spectrum peaks of Ti 2p orbits are distributed asymmetrically. This is because Ti has an unfilled d orbit. The Ti 2p energy levels are divided into Ti $2p^{3/2}$ and Ti $2p^{1/2}$ under the effect of electron spin–orbit coupling. Ti $2p^{3/2}$ characteristic peak at 458.38 eV and Ti $2p^{1/2}$ characteristic peak at 464.08 eV. The characteristic peak of Ti $2p^{3/2}$ corresponds to Ti^{3+} ions in TiN, and its half-value width is narrow, while the characteristic peak of Ti $2p^{1/2}$ corresponds to Ti^{4+} ions in TiO_2, indicating that the surface of the sample is oxidized. In order to further analyze

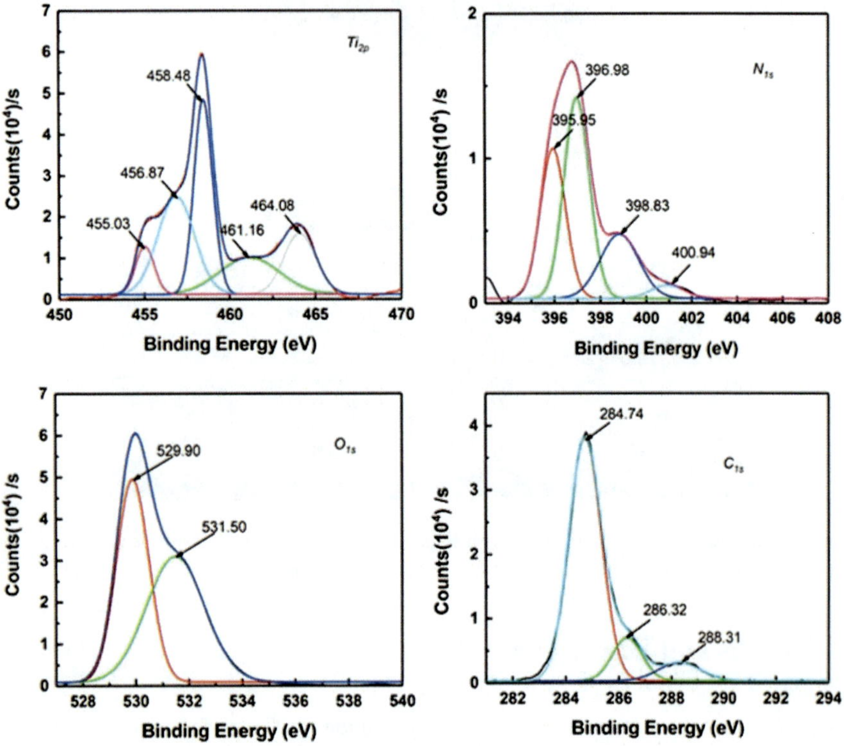

Fig. 2 X-ray photoelectron spectra of various elements in TiN films. **a** Ti, **b** N, **c** O, **d** C

the stoichiometric ratio of Ti element and N element in TiN, the ratio of each element in the TiN thin film is calculated using the formula. The calculation shows that the stoichiometric ratio of Ti to N in the film is 3.42: 1, which indicates that there are more N losses in the film.

$$Atomoic\% = \frac{\frac{I_A}{S_A}}{\sum_i \frac{I_i}{S_i}} \times 100\% \tag{1}$$

In order to analyze the optical properties of the prepared films, the films are tested using an ellipsometry. The data obtained by the spectroscopic ellipsometer cannot directly obtain the optical properties and dielectric constants of the sample materials. A model needs to be established to fit the ellipsometric data to determine the characteristics of the sample materials. In this paper, we use Drude + Tauc Lorentz's model to implement data fitting. The Drude dispersion model is used to explain the conduction electron transport characteristics in metals, transparent conductive films, and heavily doped semiconductors. Its dielectric function equation is:

$$\widetilde{\varepsilon}(\omega)=\varepsilon(\infty)-\frac{\omega_P^2}{-\omega^2+i\Gamma_d\omega} \quad (2)$$

where ε (∞) represents the high-frequency dielectric constant, which is usually greater than or equal to 1 due to high-frequency contributions; the parameters ω_p and Γ_d have important effects on the real and imaginary parts of the dielectric constant. ω_p represents the plasma frequency, which corresponds to the photon energy position where the real part of the dielectric constant is approximately 0. As ω_p increases, the amplitude of the real and imaginary parts also increases; Γ_d represents the collision frequency, and the photon absorption capacity will it expands as Γ_d increases.

In order to describe the dielectric properties of the thin film, a Tauc Lorentz dispersion model is introduced into the fitting model. The function equation of the model is:

$$\varepsilon=\varepsilon_1+\varepsilon_2 \quad (3)$$

$$\varepsilon_2=\begin{cases}\frac{1}{E}*\frac{A*E_0*C*(E-E_g)^2}{(E^2-E_0^2)^2+C^2*E^2}(E>E_g)\\ 0(E\le E_g)\end{cases} \quad (4)$$

$$\varepsilon_1=\frac{2}{\pi}*P*\int_{E_g}^{\infty}\frac{\xi*\varepsilon_2(\xi)}{\xi^2-E^2}d\xi \quad (5)$$

where A is related to the intensity of the light absorption peak, and the amplitude of the peak will increase with the increase of A, generally $10 < A < 200$; E represents the energy required for the maximum transition probability or the energy position of the absorption peak; C is the damping coefficient related to the FWHM of the absorption peak. The higher the C is, the smaller the amplitude of the absorption peak will be.

By fitting, we find that the experimental data in the visible wavelength range agree well with the fitted curve (as shown in Fig. 3).

Figure 4 shows the spectrum of (a) the real part ε_1, the imaginary part ε_2 and (b) the refractive index n and the extinction coefficient k in the complex dielectric function. The relationship between these parameters is given by the formula $\varepsilon_1 = n^2 - k^2$ and $\varepsilon_2 = 2nk$ decides. In Fig. 4, we can observe that the TiN sample has a relatively strong plasma oscillation at high photon energy; the refractive index n of the 16 nm TiN sample prepared in pure Ar environment and doped with 70% N_2 in the full energy spectrum curve has the similar trend. The value of n in the pure Ar environment is slightly greater than the value of n in the 70% N_2 environment. At high photon energy, plasma oscillation absorption is the main factor. As the photon energy decreases, low-energy photon cannot inspire ground-state electrons to jump over the forbidden band to generate excitons, and some photons fall into impurity defects. The photon absorption in the sample gradually changed from plasma oscillation absorption to other light absorption such as exciton absorption and impurity absorption. The photon absorption of TiN prepared in the 70% N_2 environment is weaker than that of the

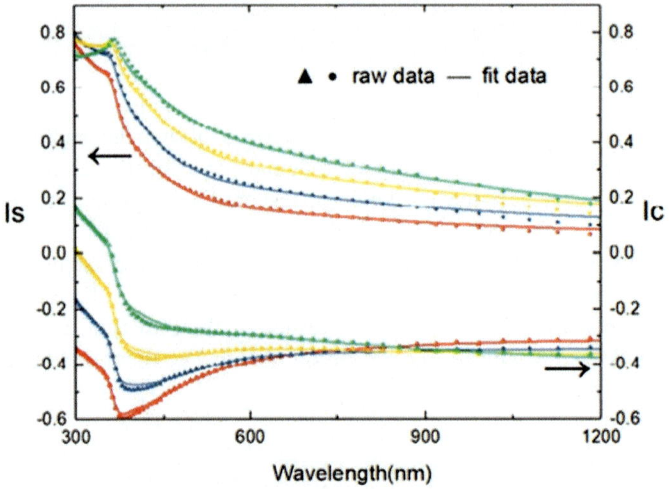

Fig. 3 Ellipsometry data fitting results of Drude + Tauc-Lorentz model

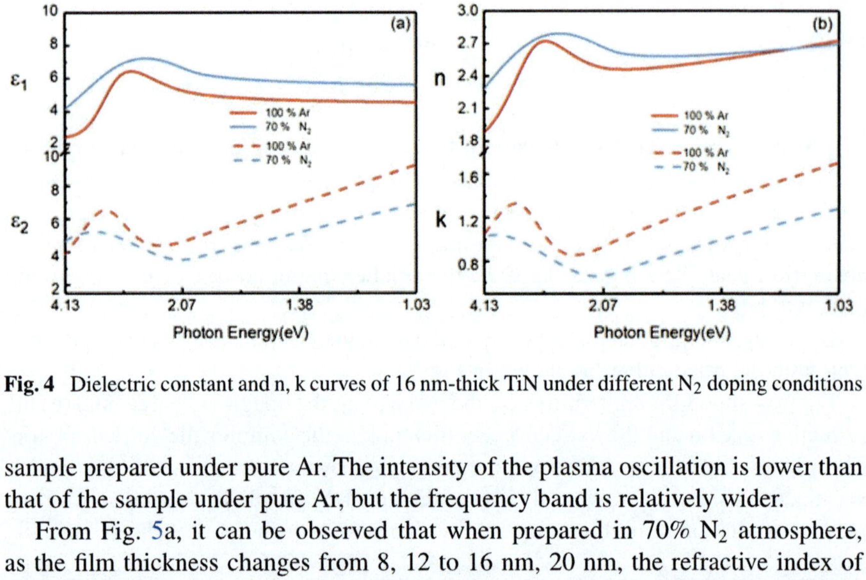

Fig. 4 Dielectric constant and n, k curves of 16 nm-thick TiN under different N_2 doping conditions

sample prepared under pure Ar. The intensity of the plasma oscillation is lower than that of the sample under pure Ar, but the frequency band is relatively wider.

From Fig. 5a, it can be observed that when prepared in 70% N_2 atmosphere, as the film thickness changes from 8, 12 to 16 nm, 20 nm, the refractive index of TiN decreases continuously in the high energy region (visible light region); When the photon energy is less than 1.56 eV, the sample with a thickness of 20 nm has a refractive index n value greater than the sample with a thickness of 16 nm, and when the photon energy is less than 1.3 eV, the refractive index n value is higher than the other two thinner samples. In Fig. 5b, when the photon energy is greater than 2.6 eV, as the thickness increases, the light absorption of the film in the high photon energy region decreases. In the range of 1.03–2.6 eV, the light absorption of the film increases with the thickness of the film. However, the k value of the TiN film with

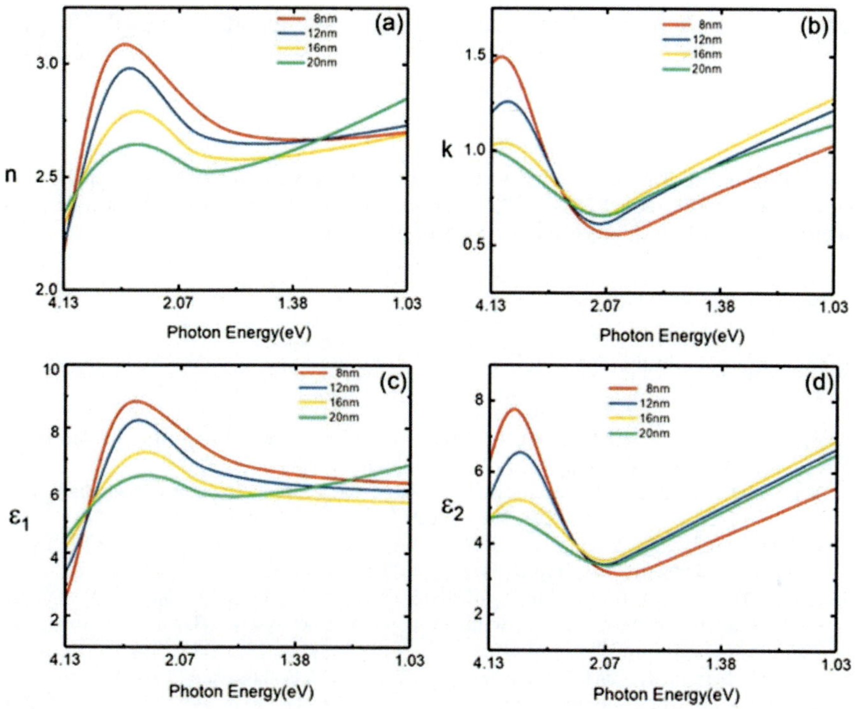

Fig. 5 TiN films prepared in N_2:Ar = 7: 3 atmosphere with a thickness of 8, 12, 16, and 20 nm

a thickness of 20 nm decreases, and its refractive index n value changes within this range. From Figs. 1b, c, it can be observed that the higher the thickness of TiN is, the larger the grain size of the surface will be; the 20 nm thickness of the sample has larger grain size compared to other samples. When the light wave propagates in the thin film, the interfacial diffuse reflection decreases, the refractive index increases, and the extinction coefficient decreases.

4 Summary

In summary, through the characterization analysis of the magnetron sputtering TiN thin film, we found that when TiN thin film was prepared using TiN as a target source, the TiN thin film prepared under argon atmosphere contained a large N deficiency. After 70% N_2 was mixed in the reaction atmosphere, the photon absorption capacity of TiN was reduced, and the refractive index was slightly increased. At the same time, when prepared in the 70% N_2 reaction environment, the extinction coefficient k value of the sample with a thickness below 20 nm will decrease as the thickness increases in the high energy region of 2.6–4.13 eV. While in the range of 1.03–2.6 eV, the k value increases with the thickness. Most of the previous reports have used Ti targets as target sources to study the optical properties of TiN thin films. This experiment hopes to provide some effective references for the preparation of devices using TiN as target sources in the future.

Acknowledgements This work was partially supported by the National Key Research and Development Program (No. 2017YFB0403501), Science and Technology Program Project of State Administration for Market Regulation (2019MK147).

References

1. Beck U, Reiners G, Urban I, Witt K (1992) Evaluation of optical properties of decorative coatings by spectroscopic ellipsometry. Thin Solid Films 220(1–2):234–240
2. Von Seefeld H, Cheung NW, Maenpaa M, Nicolet M-A (1980) Investigation of titanium-nitride layers for solar-cell contacts. IEEE Trans Electron Devices 27(4):873–876
3. Didden A, Battjes H, Machunze R, Dam B, van de Krol R (2011) Titanium nitride: a new Ohmic contact material for n-type CdS. J Appl Phys 110(3):033717
4. Hara T, Yamanoue A, Iio H, Inoue K, Washidzu G, Nakamura S (1991) Properties of titanium nitride films for barrier metal in Aluminum ohmic contact systems. Japanese J Appl Phys 30(7):1147–1151
5. Kinsey N, Ferrera M, Naik GV, Babicheva VE, Shalaev VM, Boltasseva A (2014) Experimental demonstration of titanium nitride plasmonic interconnects. Opt Express 22(10):12238–12247
6. Naik GV, Schroeder JL, Ni X, Kildishev AV, Sands TD, Boltasseva A (2012) Titanium nitride as a plasmonic material for visible and near-infrared wavelengths. Opt Mater Express 2(4):478
7. Yick S, Murdock AT, Martin PJ, Kennedy DF, Maschmeyer T, Bendavid A (2018) Tuning the plasmonic response of TiN nanoparticles synthesised by the transferred arc plasma technique. Nanoscale 10(16):7566–7574
8. Yang ZY, Chen YH, Liao BH, Chen KP (2016) Room temperature fabrication of titanium nitride thin films as plasmonic materials by high-power impulse magnetron sputtering. Opt Mater Express 6(2):540
9. Lal K, Meikap AK, Chattopadhyay SK, Chatterjee SK, Ghosh M, Baba K, Hatada R (2001) Electrical resistivity of titanium nitride thin films prepared by ion beam-assisted deposition. Physica B 307(1–4):150–157
10. Guler U, Shalaev VM, Boltasseva A (2015) Nanoparticle plasmonics: going practical with transition metal nitrides. Mater Today 18(4):227–237
11. Musschoot J, Xie Q, Deduytsche D, Van den Berghe S, Van Meirhaeghe RL, Detavernier C (2009) Atomic layer deposition of titanium nitride from TDMAT precursor. Microelectron Eng 86(1):72–77
12. Edlou SM, Simons JC, Al-Jumaily GA, Raouf NA (1994) Optical and electrical properties of reactively sputtered TiN, ZrN, and HfN thin films. In: Proceedings of SPIE - The International Society for Optical Engineering, 2262
13. Del Re M, Gouttebaron R, Dauchot JP, Hecq M (2004) Study of the optical properties of AlN/ZrN/AlN low-e coating. Surface Coat Technol 180–181(none):488–495

Analysis of Optic and Electric Combines Measurement Uncertainty Under Strong Impact Loading

Miao Zheng and Lan Wei

Abstract When the metal material undergoes a series of physical processes such as strong impact loading, unloading, reverse stretching and reloading in the complex system, the layer fracture of the material is easy to occur in the reverse stretching stage. In order to study the movement behavior and law of metal material under strong impact loading, the free surface particle velocity and the movement process of the sample surface with time were obtained by the verify of DISAR and electric probe measurement technique. In this paper, the uncertainty sources of two kinds of test systems are analyzed, and the test error ranges are given respectively. In the two dimensional Lagrange hydrodynamics program, the layer fracture processing function considering void growth, polymerization and collapse effect is added to carry out numerical simulation research. The simulation results show that the layer fracture occurs in the process of loading and unloading, and the experimental data are in good agreement with the simulation results. The width of the platform and the takeoff height of the speed curve are basically the same, and the relative error is not more than 10%.

Keywords Strong Impact Loading · Simulation · Uncertainty of Measurement

1 Introduction

When the pellet or flyer strikes the target plate, the unloading rarefaction reflected by the two free surfaces will propagate backward in the target plate [1]. The unloading wave interaction causes tensile fracture and spalling. Spallation is one of the most common failure modes of materials under impact load, and it is a typical dynamic failure mode as well as adiabatic shear fracture. The phenomenon of spallation is closely related to military engineering, aerospace engineering and other fields.

M. Zheng · L. Wei (✉)
Institute of Applied Physics and Computational Mathematics, Beijing 100094, P. R. China
e-mail: wei_lan@iapcm.ac.cn

Y. Peng and X. Dong (eds.), *Proceedings of 2019 International Conference on Optoelectronics and Measurement*, Lecture Notes in Electrical Engineering 726,
https://doi.org/10.1007/978-981-33-4110-4_17

The study of spallation also involves the interdisciplinary. Therefore, the study of spallation behavior of materials is a great theoretical and practical significance [2, 3].

At the beginning of the twentieth century, B. Hopkinson began to study the spallation behavior of metallic materials. He observed the spalling phenomenon of low carbon steel and predicted that it was most likely caused by the reflected tensile wave propagating in the material, and noted that the failure mode showed the characteristics of brittle fracture. L. Davison and A. L. Stevens, for the first time, combined mechanics with material science and put forward a preliminary physical explanation of spallation from a microscopic point of view. In China, Xing and Bai introduced statistical description method to study spallation, which promoted the development of material fracture theory. The research on the dynamic damage evolution behavior of metals generally starts from the typical dynamic tensile fracture spalling problem. However, due to the lack of systematic understanding of evolution image and physical process of spallation, there are great challenges in the interpretation of experimental data and theoretical modeling [4–6].

In this paper, we systematically analyzed the motion law of flying under strong impact loading by combining the experimental and numerical simulation work. The uncertain sources of experiment were analyzed and the range was given.

2 Experimental Facilities

Figure 1 shows the schematic diagram of tungsten flyer driven by strong detonation. The free surface velocity of the flyer was measured by DISAR, and the time when the flyer reached the fixed position was measured by electric probe.

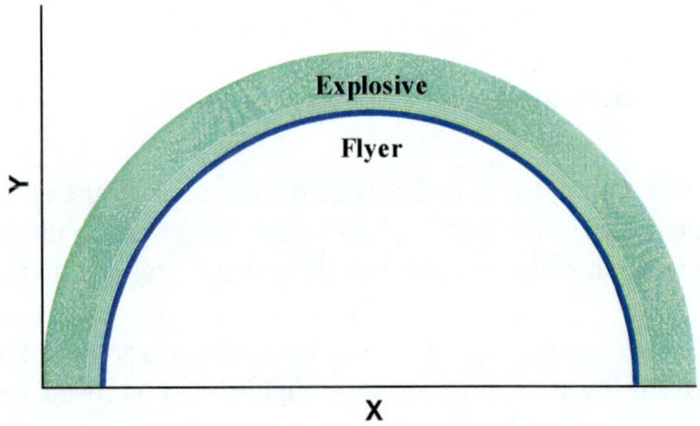

Fig. 1 Schematic diagram of experimental facility

3 Experimental Result and Uncertainty Analysis

Through two sets of test systems, the movement characteristics of flyer were verified and analyzed.

There are two sources of experimental uncertainties.

(1) The uncertainty introduced by random effect.
(2) The uncertainty introduced by system effect.

Through the analysis of this type of experiment, the uncertainty of experimental measurement mainly includes two parts. The evaluation of uncertainty is given below.

3.1 *Initial State of Experimental Device and Test System Error*

For a complex detonation system, it is difficult to give the direct relationship between the initial state variation and the measurement results. Therefore, in the electric probe test, the test points on the same test interface are regarded as multiple measurements of a single test point. The experimental standard deviation of measuring points can be calculated according to Bessel formula (1):

$$s(x_k) = \sqrt{\frac{\sum_{i=1}^{n} (x_i - \bar{x})^2}{n - 1}} \tag{1}$$

Table 1 shows the dispersion data of each interface. In order to achieve the envelope of all data, the maximum standard deviation u_e ($u_e = 77.6$ ns) of the experiment is used as the dispersion index of the electrical measurement system.

The electric and optical data in the same experiment are self consistent in theory. The integrated optical data is compared with the electrical data (Table 2).

Table 1 Dispersion of each interface

Position (mm)	40	50	70	80	90	100
experimental standard deviation (ns)	69.5	76.1	55.6	65.7	63.3	77.6

Table 2 Deviation of different systems

Position (mm)	40	50	70	80	90	100
Deviation of different systems (ns)	154	162	118	115	190	132

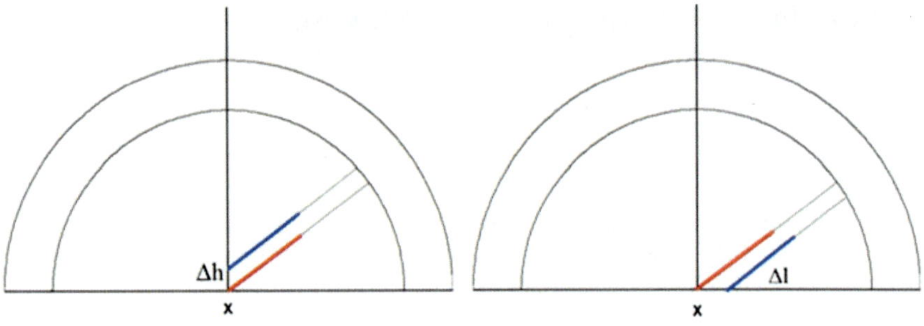

Fig. 2 Difference of measuring point position

In order to evaluate the uncertainty of optical data quantitatively, the uncertainty range of 3% is given. Within the uncertainty of the test, the optical and electric data are basically consistent.

3.2 *Difference of Measuring Point Position*

The position deviation of the measuring point is caused by the probe positioning in the testing system. Figure 2 shows the deviation caused by probe positioning height and coaxiality. The uncertainty of the interface(r) is 20 ns.

It can be seen from the data that the dispersion of the electrical measurement data is 4 times of the uncertainty of the probe positioning, which indicates that the test accuracy is high. The dispersion of the electrical measurement data mainly comes from the physical problems introduced by the structure of the device itself.

The synthetic uncertainty of the electrical measurement system is given:

$$u = \sqrt{u_1^2 + u_2^2} = 80.34 \text{ ns} \tag{2}$$

Expanded Uncertainty of electrical measurement system ($k = 2$):

$$U = ku = 160.68 \text{ ns} \tag{3}$$

The synthetic uncertainty of the optical measurement system is given:

$$u = \sqrt{u_1^2 + u_2^2} = 152.59 \text{ ns} \tag{4}$$

Expanded Uncertainty of optical measurement system ($k = 2$):

$$U = ku = 305.18 \text{ ns} \tag{5}$$

4 Numerical Simulation of Spallation

The two-dimensional Lagrangian elastoplastic hydrodynamics program integrated with VG-spall model is used to carry out numerical calculation. Figure 3 shows the comparison between the numerical simulation and the experimental data. It can be seen that the simulation results are higher than the experimental data in the first platform area of the free surface velocity curve. Generally, the height of the platform depends on the acoustic impedance and shock wave strength of the material. The dispersion of the experimental data indicates that the experimental loading conditions may not be completely consistent. The theoretical calculation shows that the free surface velocity of tungsten is about 1700 m/s, which is higher than the test value.

Table 3 shows the comparison between calculation results and electrical measurement data. Mean square deviation was used to evaluate the degree of conformity. It can be seen from the data that they are in accordance with the uncertainty of test.

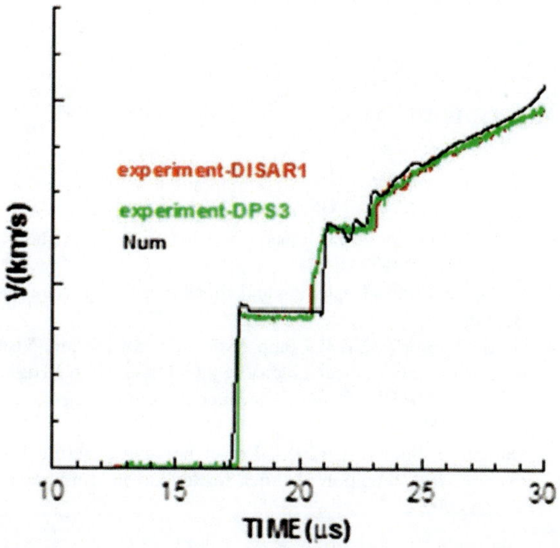

Fig. 3 Velocity curve

Table 3 Position and mean square deviation

Position (mm)	40	50	70	80	90	100
Mean square deviation (ns)	103.25	70.94	87.43	98.04	100.36	103.25

5 Analysis and Discussion

In this paper, the dynamic damage evolution characteristics of tungsten metal under the condition of strong impact loading are studied experimentally, and the free surface velocity curves of different positions are obtained. By analyzing the uncertainty sources of the experiment, the uncertainty ranges of the optical measurement system and electrical measurement system are given.

The uncertainty of optical measurement system is greater than that of electrical measurement system. This is due to the large acceleration of the target under the condition of strong primary loading, which results in the increase of the influence of the mode dispersion of the signal fiber in the laser interferometry. The results of numerical simulation can reflect the spalling phenomenon well. However, because the initial loading condition is quasi spherical wave, the three-dimensional effect of the experimental device is not considered, and there is still 10% difference between the calculation and the experimental results in the platform area. Further research will be carried out in the future.

References

1. Wang W, Wang X (2004) Research on the precursor gas accompanied with the launch of two-stage gas gun. Chin J High Pressure Phys 18(1):94–96
2. Wang X, Jia LF (2005) Broad pulse forming circuit and its application. Chin J High Pressure Phys 19(3):279–283
3. Huang ZP (2006) Explosion and shock measuring technique. National Defense Industry Press, Beijing
4. Jin S, Chen YT (2012) Comparison of multi-channel Visar and electric probe technology in measuring free-surface velocity of metal flyer. Chin J High Pressure Phys 26(5):571–576
5. Chen YT, Hu HB (2012) Experimental study of ejecta from shock melted lead. J Appl Phys 111:053509
6. Durand O, Soulard L (2013) Power law and experimential ejecta size distributions from the dynamic fragmentation of shock-loaded Cu and Sn metals under melt conditions. J Appl Phys 114:194902

Design of Fringe Projector Illumination System Based on Digital Mircomirror Device

HongYao Liang, Yan Shi, TengBiao Song, Yi Chen, Rui Xu, and ShangZhong Jin

Abstract High-speed fringe structured light projection system is one of the key technologies for real-time and efficient measurement of phase three-dimensional topography measurement system. Its function is to generate a modulated fringe pattern for three-dimensional measurement. Digital projectors based on DMD (Digital Mircomirror Device) have been widely used in three-dimensional projection fringe measurement systems due to their advantages such as high contrast, high brightness, high efficiency, and secondary development. This paper designs a high-efficiency and high-uniformity fringe projector illumination system based on DMD, with a light efficiency of 54%, a uniformity of 91.5%, and an illumination beam angle distribution of 10°–38°. The overall system size is 70 mm × 74 mm × 35 mm, which is applied to the phase three-dimensional topography measurement system, has important research significance and practical value.

Keywords Fly-eyes · DMD · Uniformity

1 Background

Since the 1990s, "optical measurement technology" has played an increasingly important role in modern industrial inspection and has become the main thrust for the development of innovative technologies in the industry. The existing reconstruction systems can be roughly divided into optical triangulation [1], laser interferometry [2], time of flight method [3], grating projection method [4, 5]. The three-dimensional contour measurement based on the grating projection method is to project a grating onto the surface of the object through an optical projector, and to capture a three-dimensional image of the grating fringe modulated by the surface of the measured

H. Liang · Y. Shi (✉) · T. Song · Y. Chen · R. Xu · S. Jin
College of Optical and Electronic Technology, China Jiliang University, Hangzhou, Zhejiang, China
e-mail: shiyan@cjlu.edu.cn

Y. Peng and X. Dong (eds.), *Proceedings of 2019 International Conference on Optoelectronics and Measurement*, Lecture Notes in Electrical Engineering 726,
https://doi.org/10.1007/978-981-33-4110-4_18

object by a camera at another position, thereby obtaining a two-dimensional deformation fringe of the grating fringe image. In the early fringe structure projection systems, gratings were mainly used. The phase shifting of the projected fringe pattern is realized by the translation of the mechanical device. The system structure is simple. With the rapid development of projection systems, the optical projector system has become a commonly used structured light projection system. The structured light pattern is programmed by a computer, so the pattern has high accuracy and strong controllability. According to different projector systems, they can be divided into cathode ray tube, liquid crystal display and digital illumination process. The cathode ray tube projection system has a large volume and a poor projection effect, and is gradually eliminated by the society. However, the liquid crystal projection system has a long response time and a low refresh frequency, which cannot meet modern measurement requirements. The DLP projection system was invented by TI. The basic component is a digital mirror. Digital mircomirror device (DMD) based digital projectors have the advantages of high brightness, high efficiency, flexible programming, and secondary development. The role of the illumination optical system in projection equipment is very important. The success of the design of the illumination optical system directly affects the performance indicators of the equipment. This article designs a fringe with compact structure, low cost and high efficiency and uniformity Projection illumination system.

2 Principle of Design

The fringe structure light projection illumination system is mainly composed of a light source, an illumination system and a DMD chip. The light from the source is homogenized by the illumination system and then modulated by the DMD chip into the projection objective, which projects the fringe pattern onto the surface of the object. Among them, the illumination system mainly includes a collimation system, a fly-eyes array and a relay system. The collimation system is used for collimating and amplifying the light beam emitted by the LED light source; the fly-eyes array homogenizes the light from collimation system to make the energy distribution uniform; the relay system is to adjust the direction of the light spot and combine the light from the fly-eyes array so that it is evenly superimposed on the DMD. This article will design a DMD-based fringe projector illumination system with an efficiency of more than 50% and a uniformity of more than 90%. The illumination beam has a numerical aperture of $F/\# = 2.4$ and the incident angle of the light in the 'ON' state is 12°. And the system is compact.

2.1 Design Initial Parameter Determination

Light source selection. In this paper, the CBT-90-B-L11 from Luminus was selected as the system light source [6], and the light source model was used to design the alignment system. The main parameters are as follows (Table 1).

DMD selection, this article chooses TI's 0.45 inch DMD display chip [7], the illumination beam is oblique incidence, the incident light requires the incident angle of the light in the 'ON' state is 12°, the numerical aperture is F / # = 2.4, the main parameters of DMD are as follows Fig. 1; Table 2).

Table 1 CBT-90 main parameters

Parameter names	Parameter values
Typical peak wavelength	456 nm
Emitting area	3.0 mm × 3.0 mm

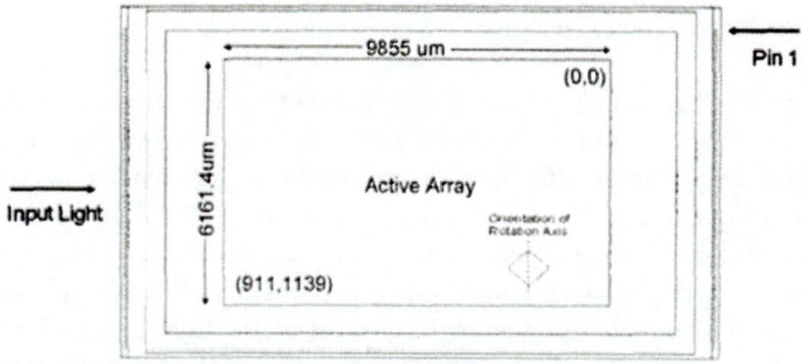

Fig. 1 0.45 inch DLP chip pixel tilt and effective size

Table 2 0.45 inch DMD chip parameter table

Parameter names	Parameter values
Pixel tilt angle	±12°
Pixel unit size	7.6 μm
Pixel arrangement method	Diagonal diamond arrangement
Resolution	912×1140
Chip length and width	9.855 mm × 6.1614 mm (16:10)
DMD efficiency	68%

Fig. 2 Collimated light path

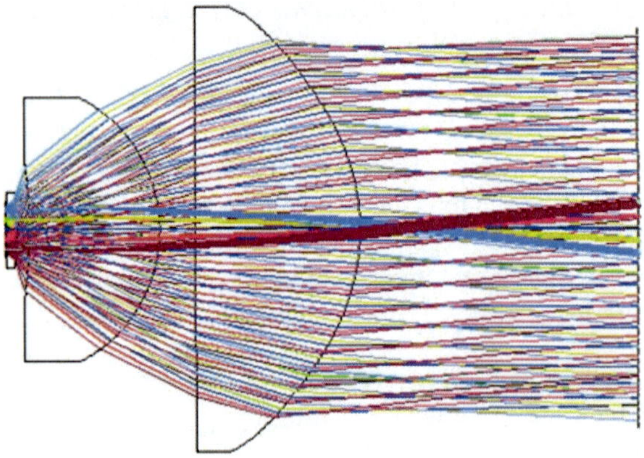

2.2 *Illumination System Design*

2.2.1 Illumination Lens Design

The fly-eyes lens has a high degree of collimation requirements for incident light. Generally, the angle of collimated light needs to be controlled within 10° to ensure high light efficiency through the fly-eyes lens. In this paper, two lenses are used for collimation design, in which the fourth surface is aspherical and the other surface is spherical or planar, which is optimized in zemax, and finally the collimated optical path as shown in Fig. 2 is obtained, which reaches the effective aperture of the fly-eyes. The efficiency is that the angle of most of the light energy is controlled within 10°. More than 90% of the emitted light energy can be utilized by subsequent systems.

2.2.2 Calculation of Fly-Eyes Parameters

A fly-eyes lens, also known as a compound eye, is a commonly used method for achieving uniform illumination. The principle of uniform illumination of the fly-eyes lens is as shown in Fig. 3. The parallel light beams are focused on the corresponding second lens after passing through the first lens, and then superimposed on the DMD through the converging lens. Because the first row of fly-eyes lenses divides the entire wide beam of the light source into multiple thin beams, and the small unevenness in each thin beam range is due to the superposition of the thin beams in symmetrical positions, the small unevenness of the thin beams is obtained Compensation, so that the light energy in the entire aperture can be effectively and uniformly utilized.

The fly-eyes homogenizing array not only provides uniform illumination for the DMD chip, but also modulates the circular illumination spot of the light source into a rectangular spot output. The aspect ratio of the fly-eyes lens unit is the same as the aspect ratio of the DMD chip, and the focal length of the fly-eyes lens and the converging lens satisfy the following relationship:

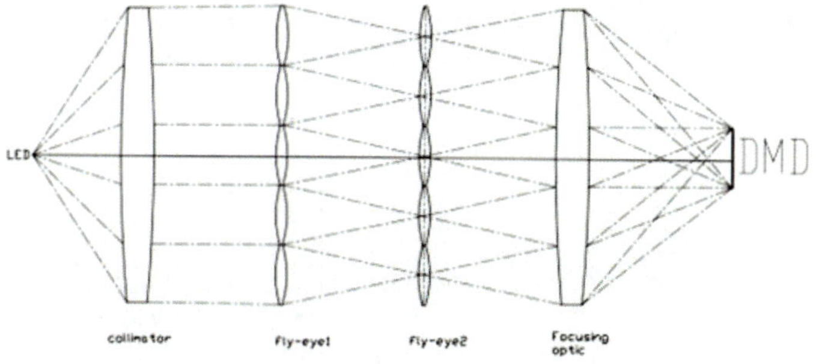

Fig. 3 Fly-eyes illumination schematic

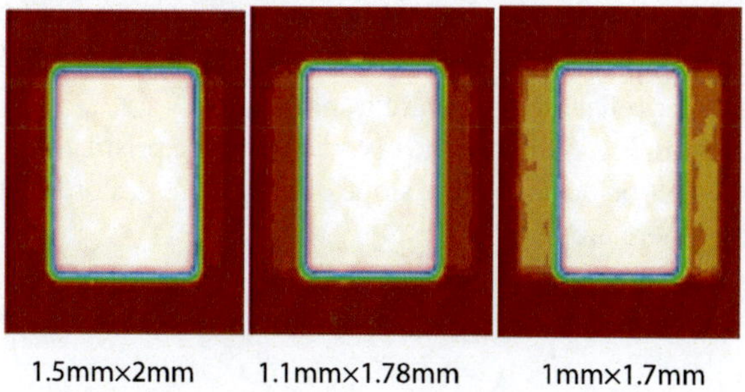

Fig. 4 Effect of Fly-eyes lens size on illumination effect

$$\frac{d}{fl} = \frac{D}{fc} \tag{1}$$

where d is the size of the fly-eyes array microlens, fl is the focal length of the microlens, D is the size of the DMD, and fc is the focal length of the focusing lens.

The 0.45-inch DMD chip used in this paper is 6.16 mm × 9.86 mm, and the LED beam is 21 mm after the beam expander system. According to the principle of the fly-eyes lens, the more the number of the fly-eyes lens unit, the better the uniform light effect. However, as the number of lens units reaches a certain level, the uniformity effect is not obvious, and side lobes are generated, as shown in Fig. 4, and the increase in the number causes the lens unit to have a very small size, is difficult to process, and the overall cost performance is lowered.

2.2.3 Relay System Design

The role of the relay system is mainly to converge the light emitted from the fly-eyes lens onto the DMD, and complete the angle requirements and direction change. The relay system includes a relay lens and an TIR prism.

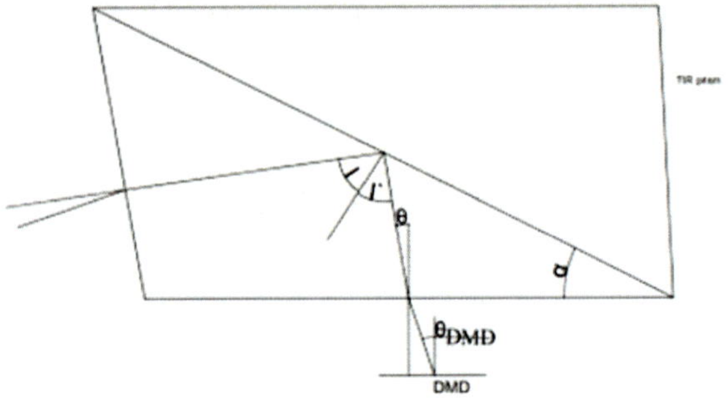

Fig. 5 TIR prism principle diagram

The principle of the TIR prism is shown in Fig. 5. It uses the principle of total reflection to select the angle of the light beam. On the total reflection surface, all the light beams that are larger than the total reflection angle are reflected, and those that are less than the total reflection angle are transmitted.

It can be seen from Fig. 5 that for the ray tracing of the prism, after the light enters the prism, total reflection occurs on the slope of the prism. According to the principle of total reflection, there are:

$$sini = sini' = \frac{1}{n} \tag{2}$$

where n is the refractive index of the prism. For light that is totally reflected by the lower surface of the prism after total reflection, the angle of the outgoing light incident on the DMD surface is θ_{DMD}. According to the law of refraction on the interface, there are:

$$\frac{\sin\theta}{\sin\theta_{DMD}} = \frac{1}{n} \tag{3}$$

At the same time, from the mathematical geometric relationship of the triangular prism:

$$\theta = i - \alpha \tag{4}$$

In this paper, $\theta_{DMD} = 12°$ (incident angle of incident light perpendicular to the DMD chip in the on state).

The fly-eyes has the same aspect ratio as the DMD, so the magnification of the relay lens group is equal to the ratio of the DMD chip to the size of the fly-eyes, and the magnification of the relay steering system can be determined to be 5.3. The object plane of the relay system is the fly-eyes exit end face, and the image plane is the DMD chip. The use of a telecentric illumination light path can maximize the uniformity of the illumination beam at the exit end of the fly-eyes reaching the DMD chip. Depending on the size of the designed prism, the rear intercept can be set to 40 mm. According to the requirements of the illumination system on the structure of

Fig. 6 Relay system design scheme

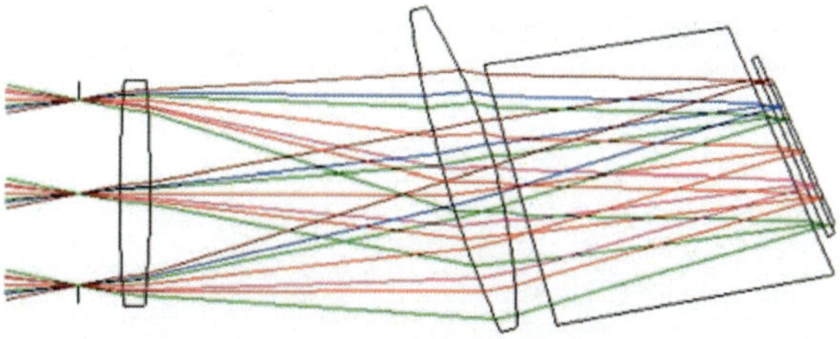

the optical machine and the back intercept, the focal length of the two lenses can be roughly obtained, so as to obtain the initial structure of the illumination light path. The TIR prism is equivalent to flat glass, which is optimized in codev to optimize the relay lens. The inclination angle, surface shape, and power distribution are used to obtain an inclined uniform illumination beam with 24° incident main surface light. The final relay system optimization results are shown in Fig. 6.

3 Simulation Analysis of Illumination System Performance

Combining the parts of the design in the previous section, the entire system is simulated in the simulation software. The light source uses a CBT-90 light file, and the receiving surface is a DMD-sized rectangular area. The optical path simulation light path diagram is shown in Fig. 7. The light spot in the DMD effective area is shown in Fig. 9, and the illuminance distribution line chart is shown in Fig. 8. The energy utilization rate at the image plane is 54%. The surface illuminance is sampled, and the uniformity is calculated to be 91.5%. As shown in Fig. 10, it can be seen that the incident angle of the light is roughly distributed around 20°, and most of the angle of the light range is between 12°–36°. The numerical aperture is ≈2.4. The overall size of the system is 70 mm × 74 mm × 35 mm. The simulation results basically meet the design requirements.

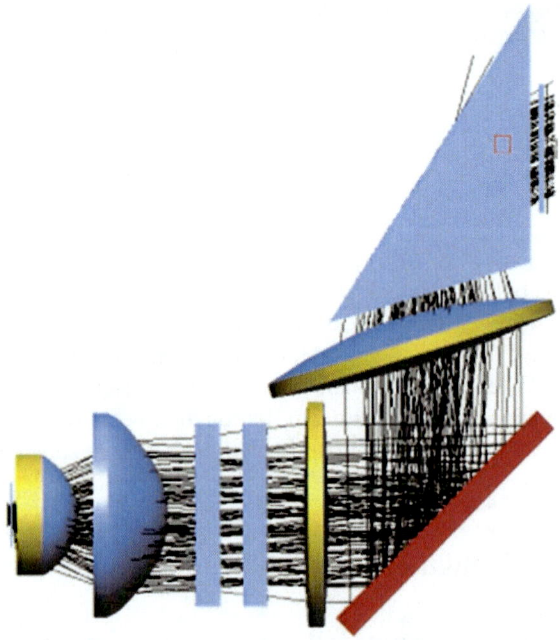

Fig. 7 Illumination system simulation

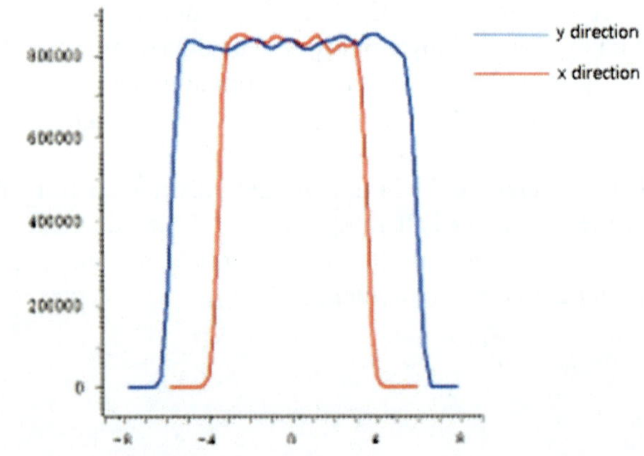

Fig. 8 DMD illumination distribution chart

4 Conclusion

In this paper, a illumination system based on DMD fringe structure is designed, the LED light source is collimated by two collimating lenses so that the divergence angle of the outgoing light beam is less than 10°. Optimize the relay lens tilt angle, power distribution, and surface shape in code, and obtain a tilted uniform illumination beam with the main light incident at 24° in the 'ON' state. The entire

Fig. 9 DMD area light spot diagram

lighting system is established in the simulation software. After testing, the light efficiency is 54%, Uniformity 91.5%, illumination beam angle distribution between 10°–38°, and the overall size of the system is 70 mm × 74 mm × 35 mm. It has important research significance and practical value for its application to phase three-dimensional topography measurement system.

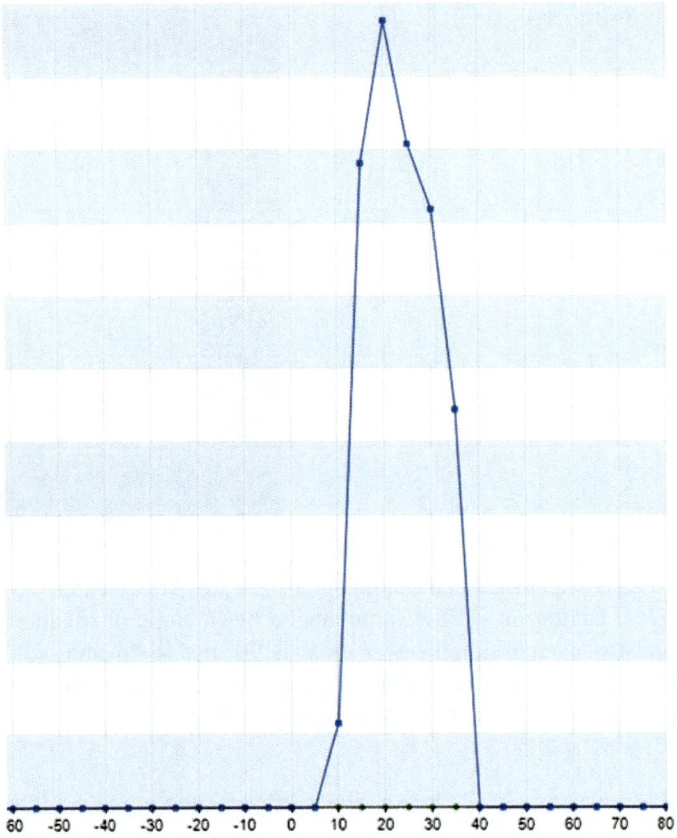

Fig. 10 DMD incident light angle distribution

Funding Major Scientific Research Project of Zhejiang Lab (2019DEKF01).

References

1. Satio K, Miyoshi T (1991) Non-contact 3-D digitizing and machining system for free-form surface. Ann CIRP, 483–486
2. Hosni Y, Ferreira L (1994) Laser based system for reverse engineering. Comput Indust Eng, 387–394
3. Chen F, Brown G, Song M (2016) Overview of 3-D shape measurement using optical methods. Opt Eng, 10–23
4. Zuo C, Huang L, Zhang M et al (2016) Temporal phase unwrapping algorithms for fringe projection profilometry: a comparative review. Opt Laser Eng, 84–103
5. Srinivasan V, Liu H, Halioua M. (1984) Automated phase-measuring profilometry of 3-D diffuse objects. Appl Opt, 3105–3108
6. Product from Luminus. https://www.luminus.com/
7. Datasheet for DLP4500 .45Wwxga DMD. https://www.dlp.com

Focusing Characteristics of Radially Polarized Anomalous Vortex Beams

Hongjie Ma, Jinsong Li, and Pengjie Sun

Abstract According to the vector diffraction theory, focusing characteristics of radially polarized anomalous vortex beams (RPAVBs) passing through a high numerical aperture (NA) lens system are studied. Mathematical formulas for the transmission process are derived. The influence of the beam order n, the topological charge m, and NA of the RPAVBs on intensity distribution is investigated in detail. Furthermore, results show that a flat-topped beam can be observed with the appropriate values of m and NA. The results are of great significance to explore the potential applications of RPAVBs.

Keywords Anomalous vortex beam · Radially polarized · Focusing characteristics

1 Introduction

The vortex beam with a spiral phase factor has attracted the attention of many scientists in the past few decades due to its wide applications in laser transmission and communication [1] and optical manipulation [2]. A completely new kind of vortex beam was proposed in 2013, which is called anomalous vortex beam (AVB). Different from the transmission properties of a normal vortex beam, this beam will eventually evolve into an elegant Laguerre-Gaussian beam in the far field (or focal plane) in free space [3]. Since the anomalous vortex beam was proposed, many scholars have studied its properties [4–6]. Yong gen Xu et al. studied the focusing characteristics of anomalous vortex beams through paraxial optical systems by combining Collins' equation and studied the beam quality of anomalous vortex beam by using the beam propagation coefficient; Yuan et al. explored the propagation characteristics of anomalous vortex beam when passing through the ABCD optical system with a side axis of annular aperture, analyzed the influence of relevant parameters

H. Ma · J. Li (✉) · P. Sun
China Jiliang University, HongZhou 310018, China
e-mail: lijinsong@cjlu.edu.cn

Y. Peng and X. Dong (eds.), *Proceedings of 2019 International Conference on Optoelectronics and Measurement*, Lecture Notes in Electrical Engineering 726,
https://doi.org/10.1007/978-981-33-4110-4_19

on the distribution of intensity distribution, which provides a theoretical basis for discovering the potential application of anomalous vortex beams; In addition, Z. Dai deduced the analytical formulas of the anomalous vortex beam in a strongly nonlocal nonlinear medium, and discussed its transmission characteristics in detail.

On the other hand, researching the focusing characteristics of beams focused by a high NA lens system is also important for their potential applications in high density optical data storage [7], microscopy [8] and particle trapping, etc. [9, 10]. For instance, Zhan et al. studied the properties of tightly focused vortex beams with circularly polarization [11], which shows that a strong longitudinal component can be produced.

In this paper, we mainly concentrate on the focusing properties of radially polarized anomalous vortex beams (RPAVBs) passing through a high NA lens system. The analytical expressions of the RPAVBs after focusing are derived. Then we analyze the influence of topological charge m, beam order n, and NA value on the intensity distribution of RPAVBs.

2 Theory

Under sinusoidal condition, the pupil apodization function of AVB (i.e., $r = f\sin\theta$) is as follows [3]

$$E_{n,m}(\theta,\varphi) = E_0\left(\frac{f\sin\theta}{\omega_0}\right)^{2n+|m|}\exp\left(-\frac{f^2\sin^2\theta}{\omega_0^2}\right)\exp(-im\varphi) \tag{1}$$

where $0 \le \theta \le \alpha$, α is the maximal value of NA, E_0 is a constant and ω_0 is beam waist width, n represents the beam order of the AVB, m refers to the topological charge. Then, r and φ are radial and azimuthal coordinates, respectively.

According to the vector Debye theory, when the radially polarized light is focused by a high NA lens, the electric field in the focal region of a high NA lens system can be expressed as [12, 13]:

$$\begin{aligned} e^s = \begin{bmatrix} e_x^s \\ e_y^s \\ e_z^s \end{bmatrix} &= -\frac{i}{\pi}\int_0^{\alpha}\int_0^{2\pi} E(\theta,\varphi)T(\theta)e^{ik(\rho_s\sin\theta\cos(\varphi-\phi)+z_s\cos\theta)} \\ &\times \begin{bmatrix} \cos\varphi\cos\theta \\ \sin\varphi\cos\theta \\ \sin\theta \end{bmatrix}\sin\theta d\varphi d\theta \end{aligned} \tag{2}$$

where $T(\theta)$ is the apodization function (for an aplanatic lens $T(\theta) = \sqrt{\cos\theta}$ [14]), $E(\theta,\varphi)$ is the pupil apodization function at the entrance pupil. The schematic diagram of the system is shown in Fig. 1. As can be seen from Fig. 1, k (k $= 2\pi/\lambda$) is the

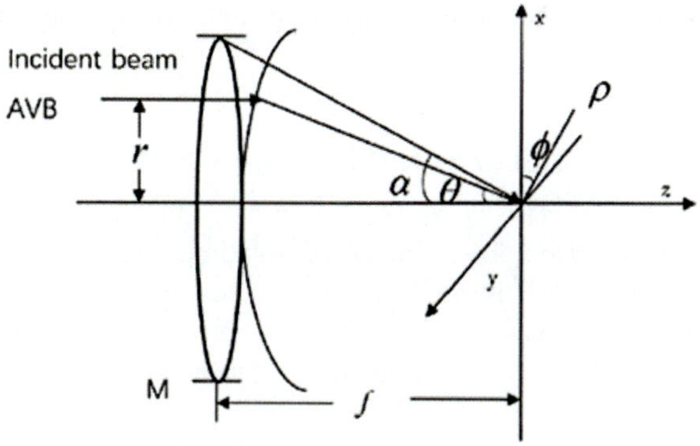

Fig. 1 Focus system diagram

wave number. The parameters ρ,φ, z are the cylindrical coordinates of the observation plane; e_x,e_y,e_z are the unit vectors along X, Y, Z directions, respectively.

Comparing with Formula (2), and using Bessel function integral formula to simplify:

$$\begin{cases} \int_0^{2\pi} \cos(n\varphi)\exp[i\alpha\cos(\varphi-\phi)]d\varphi = 2\pi i^n J_n(\alpha)\cos(n\phi) \\ \int_0^{2\pi} \sin(n\varphi)\exp[i\alpha\cos(\varphi-\phi)]d\varphi = 2\pi i^n J_n(\alpha)\sin(n\phi) \end{cases} \quad (3)$$

So, the electric field expression of the focus field the RPAVBs passing through a high numerical aperture lens is as follows:

$$E_\rho(\rho,\varphi,z) = \int_0^\alpha e^{-im\varphi}\cos\theta M(\theta) i^m [J_{m+1}(k\rho\sin\theta) - J_{m-1}(k\rho\sin\theta)]d\theta \quad (4a)$$

$$E_\varphi(\rho,\varphi,z) = \int_0^\alpha e^{-im\varphi}\cos\theta M(\theta) i^{m+1} [J_{m+1}(k\rho\sin\theta) + J_{m-1}(k\rho\sin\theta)]d\theta \quad (4b)$$

$$E_z(\rho,\varphi,z) = \int_0^\alpha e^{-im\varphi}\sin\theta M(\theta) i^m J_m(k\rho\sin\theta)d\theta \quad (4c)$$

where Fig. 4a–c, $M(\theta) = E_0\left(\frac{f\sin\theta}{\omega_0}\right)^{2n+|m|} e^{-\frac{f^2\sin^2\theta}{\omega_0^2}} e^{-ikz\cos\theta}\sqrt{\cos\theta}\sin\theta$. Focus is the origin. $\alpha = \arcsin(NA/\mathrm{n})$ is the maximum semi-convergence angle of the focusing lens. NA is numerical aperture size. $J_m(x)$ is the first type of m-order Bessel function.

Above all, we can get the total intensity of the tightly focused RPAVBs as shown below:

$$I = |E_\rho(\rho,\phi,z)|^2 + |E_\phi(\rho,\phi,z)|^2 + |E_z(\rho,\phi,z)|^2 \quad (5)$$

where E_ρ,E_ϕ, and E_z are the amplitudes of corresponding components.

3 Numerical Simulation and Analysis

In this section, using the above equations, we discuss the deep focusing characteristics of RPAVB through a high NA lens and analyzes the focusing characteristics in different values of beam order *n*, topological charge *m* and NA. The parameters that do not change in the simulation are selected as $\lambda = 632.8$ nm, $NA = 0.8, f = 2$ mm, $\omega_0 = 2$ mm.

As shown in Fig. 2, the total light intensity distribution, longitudinal and radial components of the RPAVBs with n = 1 for different m in the focal plane are presented, respectively.

It is obvious that the total light intensity distribution at the center is not zero when m = −1, 1, on the contrary, while there exists a dark spot in the center when m = 0, 2, 3. In addition, by comparing Figs. (a-1),(b-1),(c-1) and (a-3),(b-3),(c-3), we can see that the electric field of the focusing field will have the same intensity distribution when m = −1 and m = 1. To explain these phenomena, we can first find from Figs. (4) and (5) that J_m always equal to zero at the origin (except for m = 0). The first kind of Bessel function in all three components is zero at the center when m = 0, 2, 3, so the total intensity is zero. When there exists at least one component contains J_0, it means that the central intensity can be non-zero or even the maximum. Furthermore, for all three components, the focal spot size increases with the increase of topological charge m (The precondition is $m \geq 0$). And comparing Fig. 2(a-4) and (a-5), it can be concluded that the central dark spot will increase as the topological charge m increases. Above all, we can conclude that the intensity distribution of the focal field and the size of the focal spot are closely related to the value of the topological charge m.

In Fig. 3, m remains unchanged (m = 1), the total light intensity distribution, longitudinal and radial components of the RPAVBs with different beam orders n in the focal plane are shown, respectively. By comparing Figs. a(1-3), it is clear that with n increasing, the brightness of the outer rings of the total intensity distribution increased gradually, and the intensity distribution of the center pattern did not change obviously. Same situation in Figs. b(1-3) and c(1-3). So, we can draw the conclusion that the influence of the beam order n on the light intensity distribution is mainly concentrated in the brightness of the outer rings and has less effect on the central intensity distribution.

At last we study the effect of the value of NA on the focusing characteristics of RPAVBs with n = 3 and m = 1, 3. As we can see in Fig. 4, it is clear that the central intensity remains non-zero when m = 1, and when m increases to 3, the central intensity becomes dark in the focal plane.

From Fig. 4(a-1) and (a-2), it is obvious that the outer rings on the focal plane expands outward and the energy concentration of the focused light field decreases significantly with m increasing. Furthermore, comparing Fig. 4(b-1), (c-1) and (b-2),

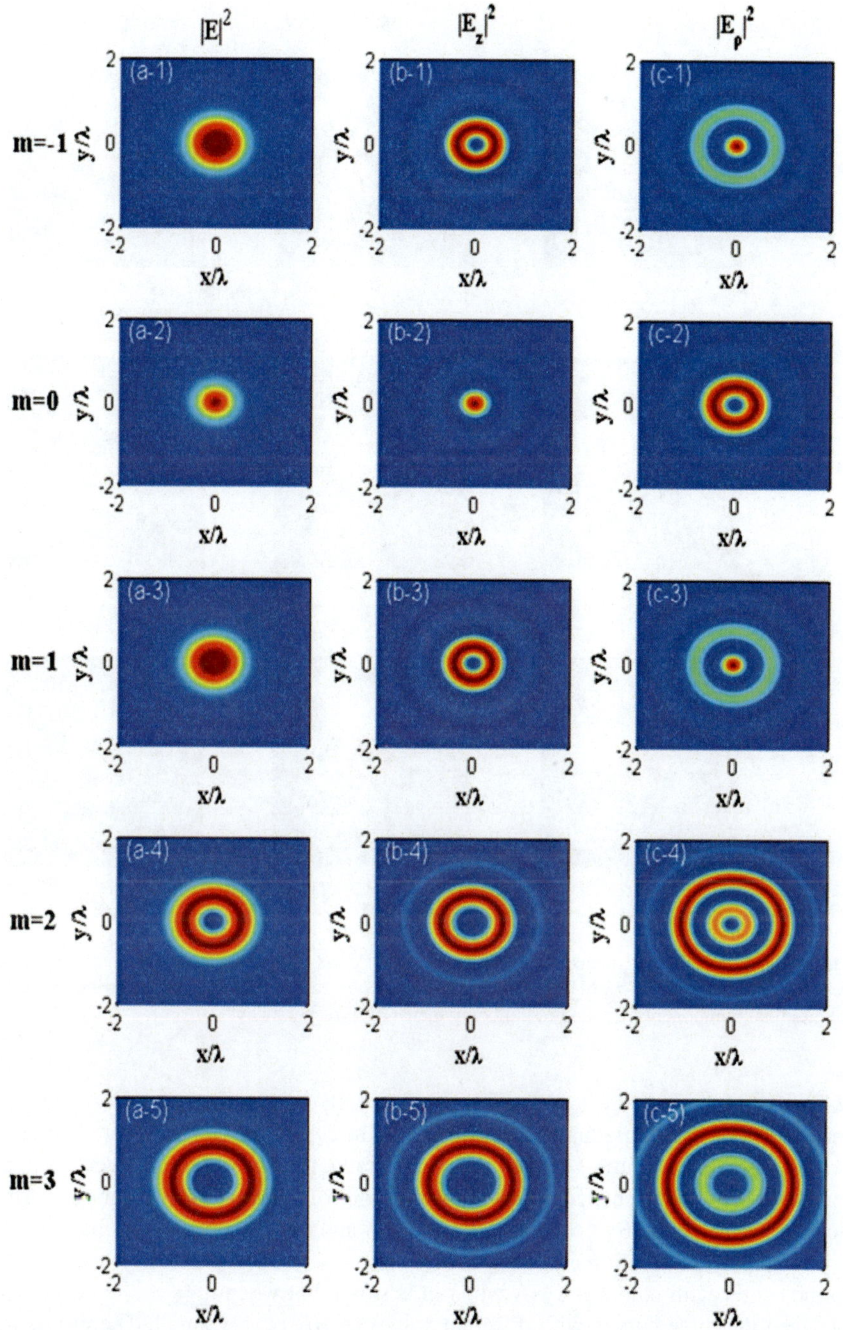

Fig. 2 Two-dimensional intensity distribution for the focused RPAVBs with $n = 1$ for different values of topological charge m. Figs. a(1-5), b(1-5) and c(1-5) correspond to the total intensity distribution $|E|^2$, longitudinal components $|E_z|^2$ and radial components $|E_\rho|^2$, respectively

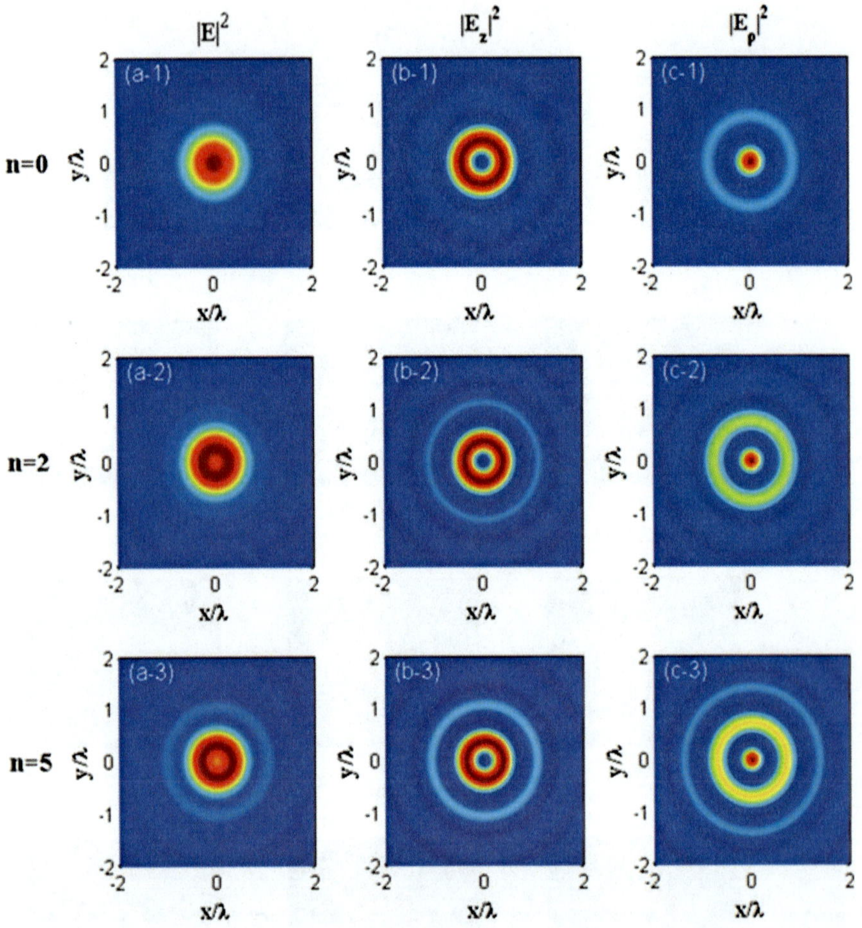

Fig. 3 Two-dimensional intensity for the focused RPAVBs with m = 1 for different values of beam orders n. Figs. a(1-5), b(1-5) and c(1-5) correspond to the total intensity distribution $|E^2|$, longitudinal components $|E_z^2|$ and radial components $|E_\rho^2|$, respectively

(c-2), it can be seen that when the value of m is small, with the increase of NA, the spot intensity distribution gradually concentrates to the center, the center pattern becomes brighter, and the brightness of the outer rings gradually increases. However, when m increases to 3, the influence of NA on the normalized distribution of light intensity decreases obviously. By comparing Fig. 4(d-1) and (d-2), we can also find that the light intensity distribution gradually evolves into a Gaussian distribution and gathers to the center with increasing NA when m is small, but when m = 3, the difference between the three curves with different values of NA is very small. Therefore, we can conclude that the influence of the NA is complicated and has a close relation with other parameters. In addition, at the appropriate value of NA and m (NA = 0.8, m = 1), a flat-topped beam can be observed (see Fig. 4(d-1)).

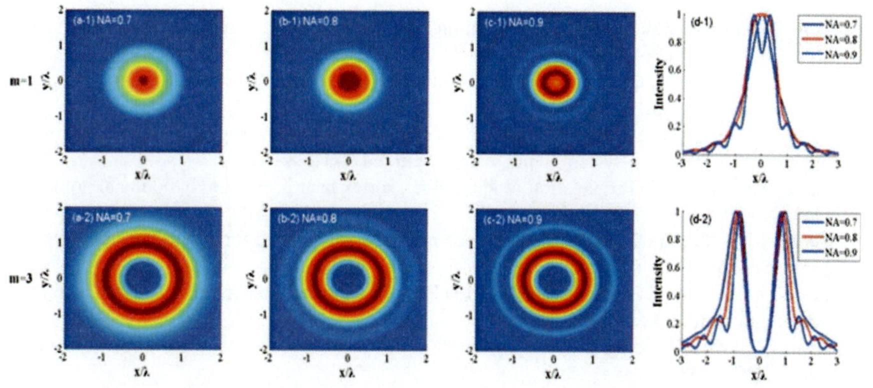

Fig. 4 Two-dimensional intensity distributions with the different NA of the RPAVBs when m = 1, 3 and n = 3. Figs. a(1-2), b(1-2) and c(1-2) correspond to NA = 0.7, 0.8, 0.9, respectively. Figs. d(1-2) are intensity cross-sections

4 Conclusions

In this paper, the focusing characteristics of RPAVBs through a high NA lens are studied. Mathematical formulas for the transfer process are derived. We have found that the value of the topological charge m has a significant effect on the intensity distribution of the focal point and the size of the focal spot. It is also clear that the influence of the beam order n on the light intensity distribution is mainly concentrated in the brightness of the outer rings and has less effect on the central intensity distribution. What is more, the results show that with NA increasing, the intensity distribution of the light spot gradually concentrate to the center, the outer rings begin to appear and the brightness continues to increase when the value of m is small. However, when m is larger, the influence of NA on beam intensity distribution becomes smaller. On the other hand, the influence of the NA is more complicated and has a close relation with other parameters (such as m). At last, it is interesting to find that we can get a flat-topped beam with the appropriate values of NA and topological charge m. The results will be helpful to the application and development of RPAVBs.

Acknowledgements This work is supported by the Commonweal Project from Science Technology Department of Zhejiang Province (No. 2017C33005).

References

1. Wang J, Yang J-Y, Fazal IM (2012) Terabit free-space data transmission employing orbital angular momentum multiplexing. Nat Photon 6:488–496
2. Dholakia K, Cizmar T (2011) Shaping the future of manipulation. Nat Photon 5(6):335–342

3. Yang Y, Dong Y, Zhao C (2013) Generation and propagation of an anomalous vortex beam. Opt Lett 38(24):5418–5421
4. Xu Y, Wang S (2014) Characteristic study of anomalous vortex beam through a paraxial optical system. Opt Commun 331:32–38
5. Yuan Y, Yang Y (2015) Propagation of anomalous vortex beams through an annular apertured paraxial ABCD optical system. Opt Quant Electron 47(7):2289–2297
6. Dai Z, Yang Z (2015) Propagation of anomalous vortex beams in strongly nonlocal nonlinear media. Opt Commun 350:19–27
7. Gu M, Li X, Cao Y (2014) Optical storage arrays: a perspective for future big data storage. Light: Sci Appl 3(5):e177
8. Lou K, Qian S-X, S., Ren Z-C, Tu C (2013) Femtosecond laser processing by using patterned vector optical fields. Sci Rep 3 (2281)
9. Zhan Q (2004) Trapping metallic Rayleigh particles with radial polarization. Opt Express 12(15):3377–3382
10. Kozawa Y, Sato S (2010) Optical trapping of micrometer-sized dielectric particles by cylindrical vector beams. Op. Express 18(10):10828–10833
11. Zhan QW (2007) Properties of circularly polarized vortex beams. Opt Lett 31(7):867–869
12. Youngworth KS, Brown TG (2000) Focusing of high numerical aperture cylindrical vector beams. OptExpress 7(2):77–87
13. Gu M (2000) Advanced optical imaging theory. Springer, Berlin, Heidelberg
14. Chen B, Zhang Z, Pu J (2009) Tight focusing of partially coherent and circularly polarized vortex beams. J Opt Soc Am A 26(4):862–869

Hybrid 2D/3D Perovskite Film with Enhanced Crystallinity via PEAI Passivation in Perovskite Solar Cells

Xin Rong, Xin Yao, and Zugang Liu

Abstract High quality and uniform perovskite films are fundamental for the performance of planar perovskite heterojunction solar cells. In this paper, phenylethylammonium iodide (PEAI) is adopted to construct 2D/3D hybrid perovskite film, because it can interact with 3D perovskite crystals to motivate the formation of perovskite phase. In addition, PEAI deposited on hybrid perovskite film is adopted to eliminate the surface traps and suppress the recombination. Furthermore, we found that perovskite film exhibits improved crystallinity with PEAI passivation, which is essential for charge transportation in perovskite solar cell. These results demonstrate that PEAI passivation layer is a credible means to enhance the performance of perovskite solar cells.

Keywords Perovskite solar cell · PEAI passivation layer · Enhanced crystallinity

1 Introduction

In recent years, hybrid organic–inorganic lead halide perovskite solar cells have triggered much attention ascribed to its remarkably increased efficiency [1]. The uniform perovskite films with complete surface cover is of great significance for optical absorption, charge transportation and device performance of perovskite solar cells (PSCs) [2]. Improving the film crystallinity and morphology are expected to suppress charge trapping and enhance carrier transporting. Nevertheless, it is difficult to prepare high-quality and uniform perovskite films, which are essential to charge dissociation and transportation in PSCs. Here, we present a strategy to enhance crystallinity and morphology of perovskite film. 2D PEAI is adopted to construct 2D/3D

X. Rong · X. Yao (✉) · Z. Liu (✉)
College of Optical and Electronic Technology, China Jiliang University, Hangzhou, P.R. China
e-mail: yaoxin_cjlu@cjlu.edu.cn

Z. Liu
e-mail: zgliu78@cjlu.edu.cn

Y. Peng and X. Dong (eds.), *Proceedings of 2019 International Conference on Optoelectronics and Measurement*, Lecture Notes in Electrical Engineering 726,
https://doi.org/10.1007/978-981-33-4110-4_20

hybrid perovskite film and utilized as the passivation layer to suppress interfacial recombination and enhance the crystallinity of perovskite films [3]. The perovskite solar cell is therefore obtained with improved performance.

2 Experiment

2.1 Materials

In this experiment, FAI (Shanghai MaterWin New Materials Co., Ltd), PbI_2 (Xi'an Polymer Light Technology Corp), CsI (Alfa Aesar, 99.999%), spiro-OMeTAD, 4-tert-butylpyridine(t-BP)(aladdin company), lithium-bis (tri-fluoromethanesulfonyl) Imie (Li-TFSI),N,N-Dimethylformamide (DMF, $\geq$ 99.9%, aladdincompany), N-Methyl-2- pyrrolidone (NMP, Sigma-Aldrich, anhydrous, 99.5%), phenethylamine (Aldrich, > 99%), ghydroiodic acid (57 wt % in H2O, Sigma-Aldrich, 99.99%) were of analytical grade and used without further purification.

2.2 Characterization

Crystallinity properties of perovskite film was recorded by X-ray diffraction (XRD) spectra (D2 PHASER) with Cu Ka radiation ($\lambda = 1.5405$ Å). The surface morphology was checked with a scanning electron microscope (SEM) (SUF-8010). While UV–visible-NIR spectrophotometer (UV-3600) was carried out to examine the absorption spectra in the wavelength of 300–900 nm. Photocurrent density–voltage (*J-V*) curves of PSCs were tested at a standard condition with light intensity of AM 1.5 G (100 mW/cm^2) and temperature of 25 °C. A metal mask of 0.1 cm^2 was clipped on ITO side to ensure the active area of PSCs with scanning rate of 35 mV/s and delay time of 50 ms.

2.3 Experiment Steps

Synthesis of phenylethylammonium iodide.

9.6 g phenethylamine was dissolved into 30 ml methanol in iced bath. 21.6 g ghydroiodic acid was slowly added into the solution. The obtained solids were washed twice with ethyl ether by a magnetic agitator overnight until the solids turned white and then it dried under vacuum to produce a white plate-like solid [4].

Device fabrication.

PSCs were fabricated with the structure of ITO glass/SnO_2/perovskite/Spiro-OMeTAD/Ag. A transparent conducting indium-doped SnO_2 coated glass substrate (ITO) was sequentially cleaned by deionized water, acetone, anhydrous ethanol and acetone isopropanol for 10 min, respectively. After dried by nitrogen gun, the ITO was dealt with plasma for 20 min. Subsequently, 60 ul of SnO_2 colloidal solution (15 wt%) was spin-coated on the ITO glasses with speed of 4000 rpm and then annealed at 150 °C for 20 min in air condition.

The substrate was then treated with plasma before transferring into the glove box. The 3D perovskite precursor solution was prepared with FAI, CsI and PbI_2 with a ratio of 0.996: 0.204: 1 in 700 mg anhydrous DMF and 125 mg NMP [5]. 2D/3D hybrid perovskite precursor solution contained a ratio of 0.9466: 0.0334: 0.02: 1 among FAI, PEAI, CsI and PbI_2 correspondingly [6].The precursor solution were both spin casted with a speed of 4000 rpm for 30 s with 150 uL chlorobenzene dripped onto the rotating substrate 20 s prior to the finish of the procedure. For the passivation layer, 8.0 mg/mL PEAI/ isopropanol solution was following spin-casted on preheated 3D and 2D/3D perovskite film respectively with the speed of 4000 rpm for 30 s. Perovskite layers were all annealed at 80 °C for 2 min at first, 150 °C for 20 min subsequently. As for the organic hole transporting layer (HTL), 80 mg Spiro-OMeTAD, 29 uL 4-tert-butylpyridine (TBP) and 18 uL LiTFSI solution (520 mg Li-TFSI in 1 mL acetonitrile) were dissolved into 1 mL chlorobenzene. The organic HTL layer was deposited at 4000 rpm for 40 s without annealing. Finally, silver electrode was thermally evaporated under a high vacuum using an Ebeam-Thermal evaporation system through a shadow mask. All the above operations are completed in the nitrogen-filled glove box.

3 Results and Disscusstion

Planar heterojunction PSCs are fabricated with 3D and 2D/3D hybrid perovskite, respectively. XRD spectrum of perovskite film in Fig. 1a demonstrates a typical diffraction pattern of trigonal perovskite phase. The stronger diffraction peaks of 2D/3D hybrid perovskite layer indicate that the crystalline quality of perovskite becomes better with PEAI doping. The PEAI can interact with 3D perovskite crystals, which is helpful for the formation of perovskite phase, making the grain boundaries functional after crystallization. The scanning electron microscopic (SEM) images in Fig. 1b depict the changes in morphology of the as-prepared perovskite films. A uniform and flat film has been achieved with the 2D/3D hybrid perovskite, with crystalline characteristics on hundreds of nanometers, eliminating the shunting and current leakage caused by uncovered 3D perovskite.

Passivation layer is further introduced to perovskite solar cell here with fabrication processes schematically depicted in Fig. 2a and PEAI is capped on perovskite surface to inhabit the surface recombination [7, 8]. The XRD spectrum in Fig. 2b illustrates dramatic improvement of crystalline quality with the PEAI layer. The photocurrent density voltage (*J-V*) curves and detailed photovoltaic parameters of these devices

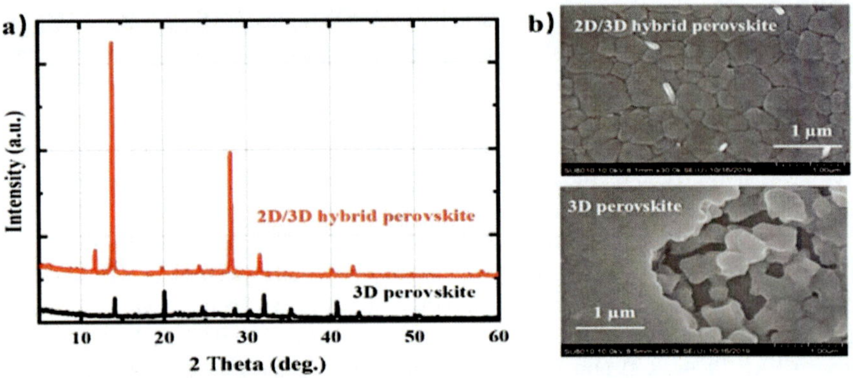

Fig. 1 **a** XRD and **b** SEM images of 3D and 2D/3D hybrid perovskite

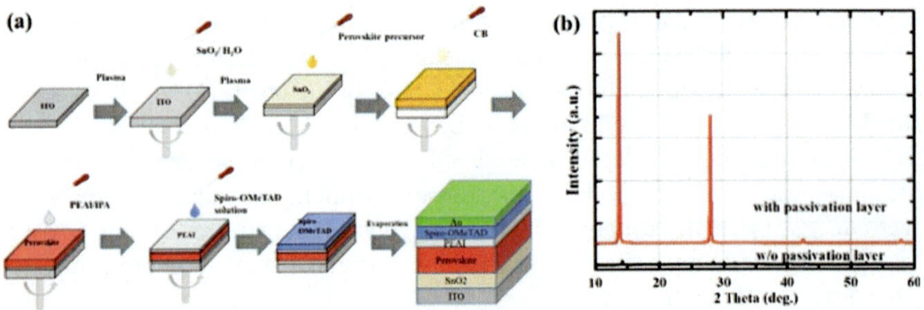

Fig. 2 **a** Schematic descriptions of device fabrication process with PEAI passivation layer. **b** XRD images of 2D/3D mixed perovskite with and w/o PEAI passivation layer

are shown in Fig. 3 and Table 1, respectively. V_{OC}, J_{SC} and FF are significantly improved with the passivation layer. The increased V_{OC} and FF result from the passivated surface recombination. While the improvement of J_{SC} can be ascribed to the better crystalline quality. Consequently, the perovskite solar cell shows the best power conversion efficiency with PEAI layer and its V_{OC}, J_{SC}, FF and PCE are 1 V, 14.8 mA/cm^2, 0.56, 8.34% respectively.

The stability is an important reflection of device performance. We have checked the device performance on the device with PEAI layer after stored for 10 days in glove box, as shown in Fig. 3b and Table 1. The PSC exhibits a stable output with ~3.6% efficiency raise after storage. The stability can be explained by a better passivation at the interface and the highly crystalline properties of the perovskite absorber [9–11].

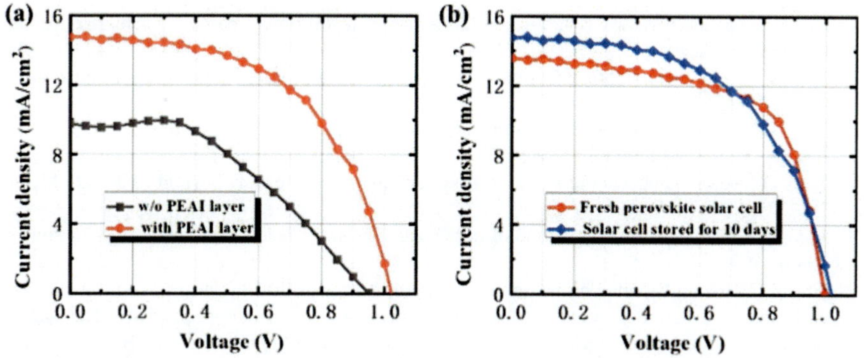

Fig. 3 Photocurrent-voltage curves of the PSC measured under simulated AM1.5 light of 100 mW cm^{-2} irradiance: **a** device with and w/o PEAI layer; **b** fresh device with PEAI layer and device stored after 10 days

Table 1 Detailed performance parameters of the PSC with and w/o PEAI layer

Sample	V_{OC} (V)	J_{SC} (mA cm^{-2})	FF	PCE (%)
Device w/o PEAI layer	0.95	10.29	0.38	3.75
Device with PEAI layer	1	14.8	0.56	8.34
Device with PEAI layer 10 days later	1	13.5	0.63	8.63

4 Conclusion

In this paper, highly uniform, dense perovskite film is successfully obtained via PEAI passivation layer. 2D/3D hybrid perovskite film is constructed through incorporating 2D PEAI into the 3D precursor solution. The hybrid film demonstrates obviously improved crystallinity and morphology. Besides, the coated PEAI further inhibits surface defect recombination and enhances perovskite crystallinity, and thus increases the *PCE*. The PSC based on PEAI layer also exhibits better stability. However the PCE of our devices are generally low, which might be caused by the damaging EB evaporation. This work gives us a new choice for preparing perovskite film with enhanced morphology and crystallinity for efficient and stable PSCs.

References

1. NREL efficiency chart. https://www.nrel.gov/pv/cell-efficiency.html
2. Tan H (2017) Efficient and stable solution-processed planar perovskite solar cells via contact passivation. Science 355:722–726
3. Cho K (2018) Selective growth of layered perovskites for stable and efficient photovoltaics. Energy Environ Sci 11:952–959

4. Lee J (2018) 2D perovskite stabilized phase-pure formamidinium perovskite solar cells. Nat Commun 9:3021
5. Jeon N (2015) Compositional engineering of perovskite materials for high performance solar cells. Nature 517:476–480
6. Lin Y (2018) Enhanced thermal stability in perovskite solar cells by assembling 2D/3D stacking structures. J Phys Chem Lett 9:654–658
7. Jiang Q (2017) Planar-structure perovskite solar cells with efficiency beyond 21%. Adv Mater 29:1703852
8. Jiang Q (2019) Surface passivation of perovskite film for efficient solar cells. Nat Photonics 13:460–466
9. Guo P (2019) Anti-colloidal-solutions: laser-generated nanocrystals in perovskite: universal embedding of ligand-free and sub-10 nm nanocrystals in solution-processed metal halide perovskite films for effectively modulated optoelectronic performance. Energy Mater 9:1970134
10. Tan H (2017) Efficient and stable solution-processed planar perovskite solar cells via contact passivation. Science 355(6326):722–726
11. Guo P (2019) Surface & grain boundary co-passivation by fluorocarbon based bifunctional molecules for perovskite solar cells with efficiency over 21%. J Mater Chem A 7:2497

Preparation and Luminescence Properties of YAG:Ce^{3+} Phosphor Based on $(Y_{1-x}Ce_x)_2O_3$ Precursor

Changhai Li, Weihe Ning, Lu Wang, Xinyue Yang, Elisha Kaseke, Ziyan Wang, Fuqiang Wang, and Hong Zhang

Abstract YAG:Ce^{3+} phosphors were synthesized by two-step method processes using $(Y_{1-x}Ce_x)_2O_3$ as a precursor. The luminescence, stability and structure of the phosphor powders were investigated by X-ray diffraction (XRD), scanning electron microscopy (SEM), and photoluminescence (PL) respectively. The XRD analysis indicates that two-step method benefits to improve phases of YAG:Ce^{3+} phosphor and obtain higher crystallinity. When Ce^{3+} doping amount was 0.67%, the two-step method was obtained maximum luminous intensity. Compared with traditional high temperature solid-state method, YAG:Ce^{3+} phosphors prepared by $(Y_{1-x}Ce_x)_2O_3$ precursor method have better thermal stability and moisture resistance. Furthermore, the packaging of YAG:Ce^{3+} phosphors was studied, which shows it can be well applied in high-power white light emitting diodes and laser lighting.

Keywords Yag:ce^{3+} · Two-step method · Phosphor · Luminescence · wLD

1 Introduction

It is widely known that white light emitting diodes (wLEDs) has many advantages in energy efficiency, environment friendly and long lifetime, compared with traditional light sources [1, 2]. Currently, commercial wLEDs are mainly fabricated by blue LED chips covering with phosphor YAG:Ce^{3+}, but blue LED chips have the "efficiency droop" problem [3], which makes LED chips unsuitable for ultra high power or brightness solid state lighting applications. Conversely, the laser diodes (LDs) have a much higher threshold for such a droop, and their efficiency increases linearly as

C. Li · W. Ning · X. Yang · Z. Wang · F. Wang · H. Zhang (✉)
College of Optical and Electronic Technology, China Jiliang University, Hangzhou, P.R. China
e-mail: zhanghong@cjlu.edu.cn

L. Wang (✉) · E. Kaseke
School of International Education, China Jiliang University, Hangzhou, P.R. China
e-mail: wanglu@cjlu.edu.cn

Y. Peng and X. Dong (eds.), *Proceedings of 2019 International Conference on Optoelectronics and Measurement*, Lecture Notes in Electrical Engineering 726,
https://doi.org/10.1007/978-981-33-4110-4_21

the current density increases [4]. In order to maximize the advantages of LDs in the process of pumping the phosphor material, the innovative high thermal phosphor is an urgent need in high-power white light emitting diodes and laser lighting.

$YAG:Ce^{3+}$ phosphors are usually prepared by high temperature solid-state method, the raw materials are fully mixed and calcined at high temperature [5, 6]. However, the problem with this method is that the grain size of the powder it produces is relatively large and contains a stable intermediate phase. This may be due to the fact that the raw material cannot be mixed completely and uniformly. A variety of improved wet chemical methods, such as the sol–gel method, co-precipitation, and solvothermal method, have been used to synthesize $YAG:Ce^{3+}$ phosphors. Even if advantages such as fine homogeneity, hyperreactivity of the raw materials, and low temperature of sintering as those methods carry, the procedure is time-consuming and involve complex reactions. Therefore, a two-step method using $(Y_{1-x}Ce_x)_2O_3$ as a precursor based on traditional high-temperature solid-state method have been studied.

In this study, $YAG:Ce^{3+}$ phosphor was prepared by two-step method using $(Y_{1-x}Ce_x)_2O_3$ as a precursor. The luminescence, structure of the phosphor powders were investigated for various sintering temperatures, different doping concentration of Ce^{3+} using X-ray diffraction (XRD), scanning electron microscopy (SEM), and photoluminescence (PL). Furthermore, it is necessary to study the optical properties at high temperature and humidity, and compare the luminescence properties of the phosphors prepared by two different preparation methods.

2 Experiment

$YAG:Ce^{3+}$ phosphor was synthesized by two-step method using $(Y_{1-x}Ce_x)_2O_3$ as a precursor. The reagents Y_2O_3(A.R.), Al_2O_3(A.R.) and CeO_2(A.R.) were used in the experiment. Firstly, grind the stoichiometric mixture of Y_2O_3(A.R.) and Al_2O_3(A.R.) in an agate mortar. then, followed by a 4 h heating at 1500°C in the tubular furnace with the atmosphere of H_2(5%) + N_2(95%). The precursor, $(Y_{1-x}Ce_x)_2O_3$ was ground and mixed with CeO_2(A.R.) in stoichiometric ratio before being heated again for 4 h at 1400°C under the same reducing atmosphere. The mixture was cooled to room temperature and the samples were grind to obtain $YAG:Ce^{3+}$ phosphor. At the same time, another $YAG:Ce^{3+}$ phosphor for comparative was prepared by a high-temperature solid-state method. A stoichiometric mixture of Y_2O_3(A.R.), Al_2O_3(A.R.), CeO_2(A.R.) was grinded in an agate mortar. Finally, load the mixtures into a tube furnace with the atmosphere of H2(5%) + N_2(95%), sintered at 1500 °C for 4 h.

3 Results and Discussion

The X-ray diffraction patterns of $YAG:Ce^{3+}$ phosphors synthesized by the high temperature solid state method and two-step method (by using $(Y_{1-X}Ce_x)_2O_3$ as a precursor) was shown in Fig. 1. In both of two preparation methods, all the diffraction peaks are consistent with the standard patterns of YAG. And the diffraction peaks of the YAG powder prepared by two-step method processes are more sharp than that at a sintering temperature of 1500°C. It shows that the crystal structure of the powder in two-step is more developed and more stable.

Figure 2a shows SEM image of the $YAG:Ce^{3+}$ phosphors prepared by high temperature solid state method, It can be seen that the diameter of the $YAG:Ce^{3+}$ phosphors powder particles ranges from 10–15 um. Figure 2b shows SEM image of

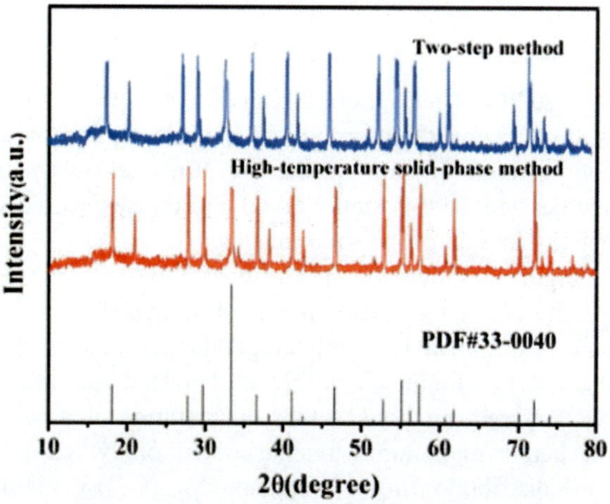

Fig. 1 X-ray diffraction patterns of the $YAG:Ce^{3+}$ samples

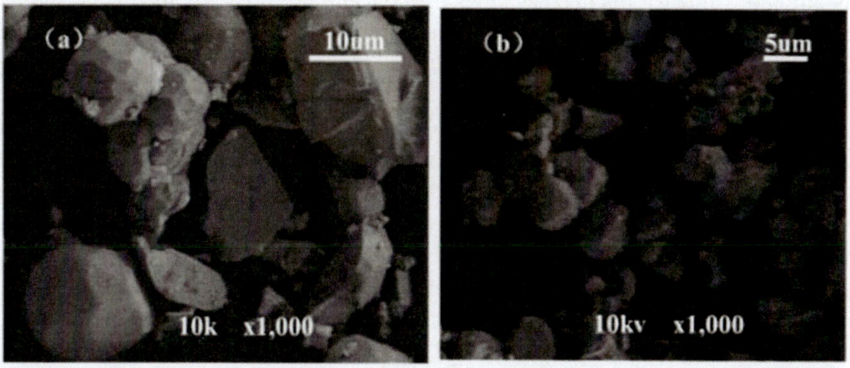

Fig. 2 SEM of samples synthesized by high temperature solid state method (**a**) and two-step method (**b**)

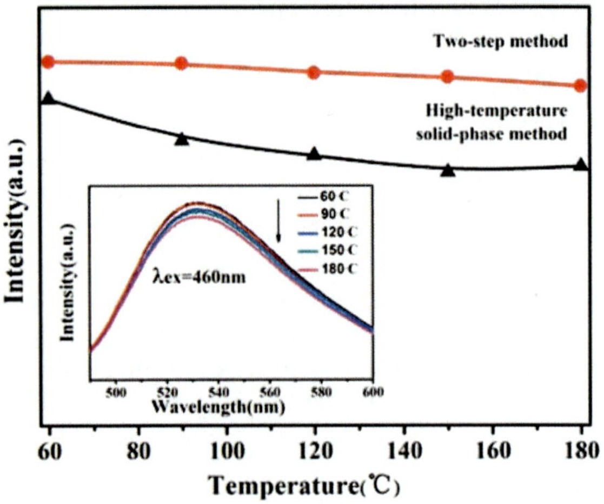

Fig. 3 The PL patterns of $YAG:Ce^{3+}$ at different Ce^{3+} concentration

the $YAG:Ce^{3+}$ phosphors during two-step processes. The diameter of the particles is about 5 um, the morphology is uniform, and the distribution of particle size is concentrated. To sum up, the $YAG:Ce^{3+}$ crystals with uniform distribution can be synthesized by two-step method and the surface characteristics of the crystals are better than that of the solid images.

Figure 3 shows the spectrum of $YAG:xCe^{3+}$ ($x = 0.02, 0.04, 0.06, 0.08, 0.1$) at different concentration prepared by two-step method and $YAG:Ce^{3+}$ with doped concentration 0.04 by high temperature solid-state method. As the figure shows that there is no significant difference between the intensity of phosphor prepared by two different methods at the concentration of x was 0.04. The luminous intensity reached its maximum when the content of x was 2%. Accounted for the fact that the Ce^{3+} distribution in the precursor $(Y_{1-x}Ce_x)_2O_3$ is more uniform, which weakens interionic interactions and improves the luminescence intensity. In two-step method, the luminous intensity is decreasing as the concentration of Ce^{3+},increases, and its best doping concentration is $x = 0.02$, indicating that the efficiency of substitution of Y^{3+} by Ce^{3+} is higher. It results in the occurrence of concentration quenching shifting towards low concentration.

The comparison of thermal stability of $YAG:Ce^{3+}$ phosphors prepared by different methods and the emission spectrum of $YAG:Ce^{3+}$ phosphors by two-step method sintered at various temperatures are shown in Fig. 4. We found that with the increase of temperature, the luminescence intensity of $YAG:Ce^{3+}$ phosphors prepared by both methods decreased, and it lead to the luminous decay. When the temperature reaches 120 °C and kept it for 8 h, the relative emission intensity of $YAG:Ce^{3+}$ phosphors synthesized via high temperature solid state method decreased by 10%, however, that decreased by 3% in two-step method. The intensity of powder decreased by 6% in two-step method using $(Y_{1-x}Ce_x)_2O_3$ as a precursor rather decreased by 14% in solid phase method when up to 180°C. It can be inferred that the two-step method is

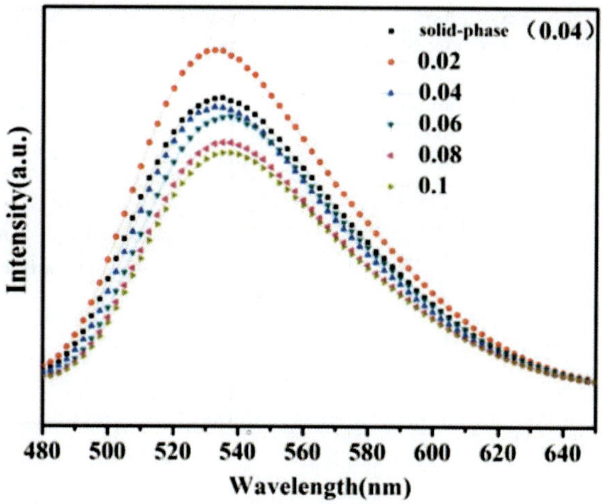

Fig. 4 Comparison of thermal aging of YAG:Ce^{3+} phosphors

a better approach to maintain the luminescence performance and thermal stability at high temperature compared with the solid phase method.

Figure 5 is the emission spectrum of YAG:Ce^{3+} phosphor prepared by different method under different humidity conditions. As shown in Fig. 5, with the increase of humidity, the luminescence intensity of the phosphor samples prepared by the two methods attenuated, which may be in a hot and humid environment, the phosphor decayed by temperature. The phosphor is dampened by water erosion, resulting in a decrease in luminescence intensity. Under the conditions of humidity controlled by 180°C, the luminescence intensity of the samples prepared by solid state method decreased by 22%, while that from two-step method using

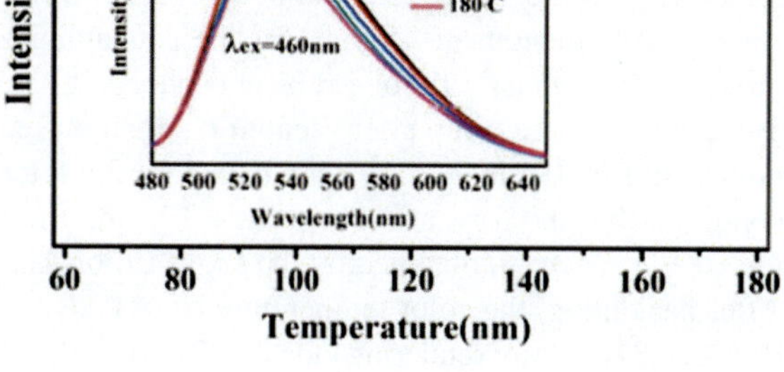

Fig. 5 Comparison of the emission spectrum of YAG:Ce^{3+} phosphors in different humidity environments

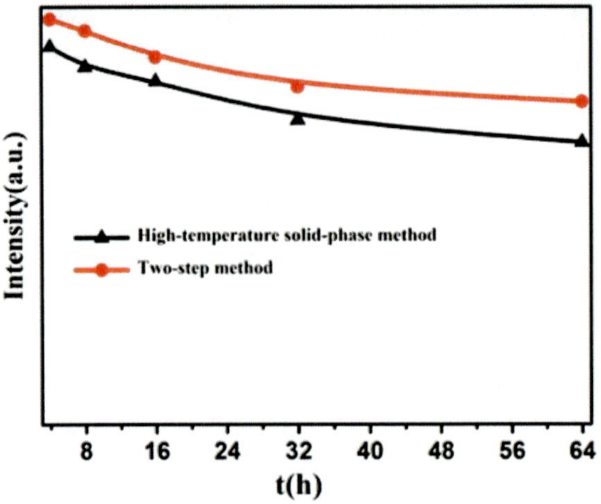

Fig. 6 Variation of luminescence intensity of YAG:Ce^{3+} prepared by under certain humidity environment with time

$(Y_{1-X}Ce_x)_2O_3$ as a precursor decreased by 18%, indicating that the YAG:Ce^{3+} phosphors prepared by high-temperature solid phase method are more susceptible to erosion by high-pressure steam than two-step method.

Figure 6 shows the variation of luminescence intensity over time of YAG:Ce^{3+} phosphors by different methods under a certain humidity condition. The luminescence intensity of the phosphors prepared by two methods was found to decrease when the samples were subjected to humidity at 180 °C for 16 h, which decreased more slightly by two-step method. When the processing time is extended to 32 h and above, the phosphors prepared by the two methods still have the phenomenon of light decay. Thus, compared with traditional solid phase method YAG:Ce^{3+} phosphors prepared by $(Y_{1-X}Ce_x)_2O_3$ precursor method have better moisture resistance, but which is limited and the powder needs to be further optimized to improve its application at high temperature and humidity.

The luminescence performance of phosphors applied to wLED is studied, and ultimately it needs to be encapsulated on the LED chip to achieve the efficient and high stability output of wLED light source. Therefore, this part studied the packaging of YAG:Ce^{3+} phosphors prepared by two-step method and blue light LED chips into white LED. The YAG:Ce^{3+} phosphor sample (538 nm) prepared when doped with 0.67% Ce^{3+} was selected to combine with the commercial blue light LED chip with emission wavelength of 452 nm for the packaging experiment of wLED. First, a mixture with a mass ratio of 1:3:3 (phosphor:S28A glue:S28B glue) was dotted on a 452 nm blue-LED chip after vacuuming, then placed in a LED photoelectric oven, preheated for 30 min at 70 °C, and baked for 3 h after heating to 150 °C. Under the condition that the forward voltage was 3 V and the forward current was 150 mA, the wLED spectrum obtained by the Light Color Analysis Test is shown in Fig. 7a. After packaging, the color temperature for wLED is 7802 K, color coordinates are (0.296, 0.310), color rendering index is 80.3, luminescence efficiency is 110.09 lm/W. The test results of wLED indicate that the powders synthesized by $(Y_{1-X}Ce_x)_2O_3$

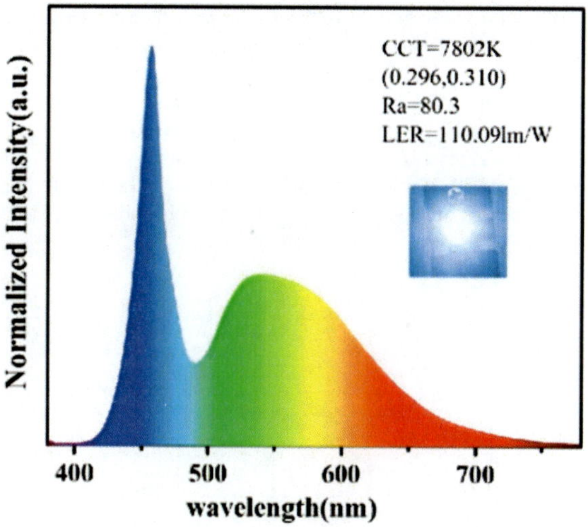

Fig. 7 Spectrum of the wLEDs based on InGaN blue-LED chip

precursor method has a promising application in cool color lighting scene, and had the ability to achieve white light source when combined with blue light chip, which is suitable for the packaging preparation of high color temperature wLED and has the practical market potential.

4 Conclusions

YAG:Ce^{3+} has been synthesized by two-step method using $(Y_{1-x}Ce_x)_2O_3$ as a precursor. The XRD pattern and SEM pattern of the samples shown the two-step method is an effective way to acquire better purity phase and more uniform morphology. The diameter of the particle is between 2 and 5 μm. Different doping concentration has a significant impact on the luminescence properties of the phosphors. The luminous intensity reached its maximum when doped with 0.67% Ce^{3+}. Compared with the solid-phase method, the two-step method based on $(Y_{1-x}Ce_x)_2O_3$ precursor was an effective method to obtain better luminous performance, thermal stability and moisture resistance. The physical and optical properties of YAG:Ce^{3+} phosphors were improved and can be well applied in high-power white light emitting diodes and laser lighting.

References

1. Wang L, Xie R, Suehiro T et al (2018) Down-Conversion nitride materials for solid state lighting: recent advances and perspectives. Chem Rev 118(4):1951–2009
2. Kitai A (2016) Materials for solid state lighting and displays. Wileyd

3. Crawford MH (2009) LEDs for solid-state lighting: performance challenges and recent advances. IEEE J Sel Top Quantum Electron 15(4):1028–1040
4. Wierer JJ, Tsao JY, Sizov DS (2013) Comparison between blue lasers and light-emitting diodes for future solid-state lighting. Laser Photonics Rev 7(6):963–993
5. Shusheng Z, Weidong Z, Chunlei Z (2004) Study on $(Y,Gd)_3(Al,Ga)_5O_{12}:Ce^{3+}$ phosphor. J Rare Earths 22(1):118–121
6. Fu S, Tan J, Bai X et al (2018) Effect of Al/Ga substitution on the structural and luminescence properties of $Y_3(Al_{1-x}Ga_x)_5O_{12}:Ce^{3+}$ phosphors. Opt Mater 75:619–625

Tunable Diode Laser Absorption Spectroscopy for Non-invasive Detection of Microbial Growth

Zhemin Chen, Pengbing Hu, Sunqiang Pan, Jianfeng Zhang, Gangxiang Guo, and Sumei Liu

Abstract This paper reports the use of tunable diode laser absorption spectroscopy (TDLAS) to measure the carbon dioxide (CO_2) in the culture vials. The accumulated CO_2 concentration in the headspace is linked with the growth of the microorganisms cultured in the vials. The 2004.1 nm spectral line of CO_2 is selected as the sensing transition for its strong absorption strength. The proposed TDLAS sensor is calibrated and verified with several CO_2 standard gases, yielding a minimum 1σ detection limit of 0.084% with an integral time of 3 s. The suitability of the sensor to the microorganisms is performed, almost the results agree with the growth laws of microbiology. Since the sensor is fast and nondestructive, it can be promising for in-line detection of microbial growth.

Keywords Carbon dioxide · TDLAS · Headspace · Microbial growth

1 Introduction

Aseptic processing is widely used in the food and pharmaceutical manufacturing [1–3]. In the aseptic environment, it is periodically verified by filling microbial culture media instead of the food or drug product to ensure the reliability and repeatability of aseptic processing. In traditional way, we often incubate the intact media-filled vial for 7–14 days and then assess the microbial contamination by visual inspection. It suffers from long time and high cost. Besides, the food and drag production also risks of microbial contamination in the packaging, transportation and storage process. Presently, many invasive techniques are utilizing to detect micro-organism growth, because of their selectivity and reliability, such as polymerase chain reaction and bacterial enumeration. It takes several hours and requires professionally trained microbiologists.

Tunable diode laser absorption spectroscopy (TDLAS) is based on gas molecule absorption characteristic at a specific wavelength range. The concentration of gas

Z. Chen · P. Hu (✉) · S. Pan · J. Zhang · G. Guo · S. Liu
Zhejiang Institute of Metrology, Hangzhou, China
e-mail: hupengbing@163.com

Y. Peng and X. Dong (eds.), *Proceedings of 2019 International Conference on Optoelectronics and Measurement*, Lecture Notes in Electrical Engineering 726,
https://doi.org/10.1007/978-981-33-4110-4_22

matter can be determined by measuring absorption strength via Lambert–Beer Law. With increasing maturity and broader availability of laser light sources, TDLAS has found use in numerous applications in food and pharmaceutical manufacturing. For leakage detection of pharmaceutical vials, Yang et al. measured leakage coefficient defined by incident and transmitted light intensities to judge the deterioration of pharmaceuticals in the vials [4]. Cai et al. proposed a TDLAS sensor for accurate measurements of commercial freeze-dried products. The referenced signal was used to reduce the influence of ambient humidity extensively [5]. A fast, invasive and in-line measurement technique was developed in the quality control of food packaging [6]. TDLAS sensor was employed for automatically headspace oxygen sensing in flow-packed products. Brueckner et al. made a comparison of TDLAS and isothermal micro-calorimetry for microbial growth detection in media fills [7]. And it showed the proposed TDLAS sensor was recommended for automating the media-fill inspection.

In this work, a TDLAS sensor has been developed for absolute measurement of the concentration of carbon dioxide in microbial culture vials. During respiration in aerobically respiring microbes, CO_2 accumulates by oxidation–reduction process in different phases of growth. Also, aero-tolerant or anaerobic bacteria can metabolize carbon dioxide. Thus, the microbial growth status can be traced by measuring CO_2 formation in the vial headspace. The performance of the sensor is firstly calibrated and validated on several CO_2 standard gases. It shows the sensor is qualified with 1σ detection limit of 0.084% ranging from 0 to 20%. Secondly, two types of bacteria, *Bacillus subtilis (B. subtilis)* and *Stapbylococcus aureus (S. aureus)*, are tested by the sensor, showing a good agreement with the microbial growth law.

2 Setups and Methods

The developed automated and non-intrusive TDLAS sensor for detecting microbial growth in the culture vials is illustrated in Fig. 1.

A tunable DFB laser emitting at 2004.1 nm was used as a light source, controlled by a laser controller. The laser beam, collimated by the optical fiber collimator, was then directed through a culture vial with a headspace path length of 3 cm. Then, the transmitted laser beam was focused on a photo detector. Here it is notable that the collimator and detector should be close to the vial's outline as far as possible. This configuration can effectively eliminate the excess gaseous matter absorption caused by external path length. In order to increase the sensitivity of the CO_2 measurement, wavelength modulation spectroscopy (WMS) techniques were adopted. Thus, the laser modulation parameters were set as follows. The laser was scanned by a saw-tooth wave at 10 Hz, overlapped with a 4 kHz high frequency sine-wave carrier. Last, the second harmonic signal of the headspace CO_2 gas absorption was demodulated by a lock-in amplifier. The setup was placed inside the incubator to provide suitable microbial growth temperature, and several culture vials were together mounted on a rotating table for automatic sample feed in advance.

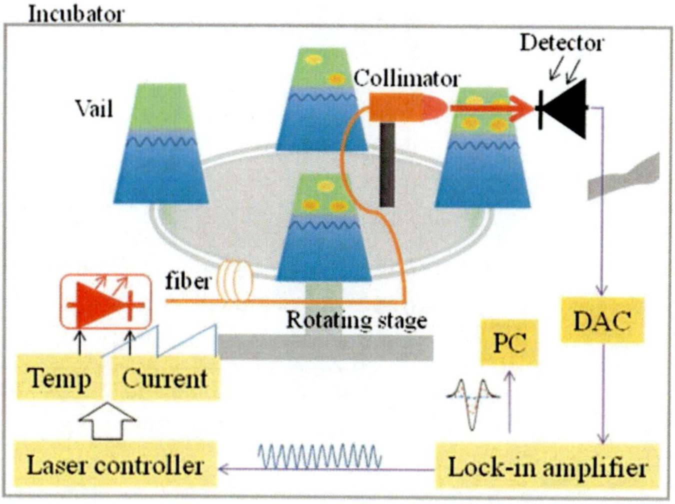

Fig. 1 Scheme of the experimental setup used to measure microbial growth

As the laser beam propagates in the vial headspace over an absorption path length of L, the beam will be absorbed in situ by the gas molecules of interest. The absorption of laser intensity is related to the molecular absorption coefficient, molecular mass concentration and absorption length, based on the Beer-Lambert law. The transmitted intensity I can be described as follows [8, 9].

$$I = I_0 exp(-\sigma CL) \quad (1)$$

where I_0 is the incident laser intensity in the absence of absorbing species, σ is the molecule absorption coefficient in standard state, as a function of ambient temperature and pressure and so on, C is the concentration of the absorbing species in molecules per unit volume and L is the optical path length. In WMS scheme, it is only concerned about weak absorption, i.e., the value of σCL is much less than 1. In this situation, Eq. (1) can be expanded in a Fourier cosine series and the second harmonic signal obtained by demodulation could be simply expressed as [10]

$$I_{2f} = I_0 \sigma CL \quad (2)$$

It reveals that the product of the incident laser intensity I_0, the gas concentration C is proportional to the calculated second harmonic signal value I_{2f}. The I_0 was simultaneously processed with I_{2f} to reduce the influence of the laser intensity fluctuation reaching the detector. The optical length L was fixed after the fabrication of the sensor and the absorption coefficient σ was also a constant in standard state, and thus the gas concentration could be able to solely and correctly derived from the equation.

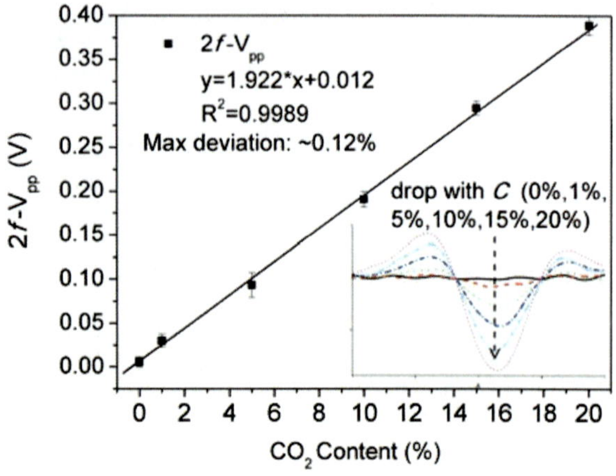

Fig. 2 Harmonic signal versus CO_2 content at different concentrations

3 Experimental and Results

The proposed sensor was firstly calibrated and validated by measuring CO_2 content in the vials flushed with several CO_2 standard gases of 1, 5, 10, 15 and 20% volume concentrations. It was carried out in a constant temperature and humidity laboratory, with the environment condition of 30 °C, 65.5% and 101.3 kPa. The inset of Fig. 2 shows the tendency of the *2f* absorption spectra at different CO_2 gas concentrations. The valley of the absorption spectrum drops gradually with the increment of the CO_2 concentration, while the peak changes in the contrary way. This means the use of the peak to valley value of the absorption spectra will gain a maximum measurement sensitivity. The peak to valley value versus concentration was plotted in the Fig. 2. It shows that the linear fitting degree is 99.98% in the measurement range from 0 to 20%. The good linear characteristics indicate the sensor can detect the CO_2 concentration inside the headspace of the culture vials.

To verify the sensor's stability, CO_2 measurements with 10% gas concentration were performed over a period of ~1 h, and the measured data were recorded as shown in Fig. 3. An Allan deviation is utilized to analyze the detection limit, and a standard deviation to the stability. The left part of Fig. 3 exhibits the Allan deviation as a function of the integral time. The plot indicates that the 1σ detection limit is 0.121% for a 0.1 s measurement time, and also shows an optimum integral time of 3 s corresponding to a minimum 1σ detection limit of ~0.084%. The measured concentration fluctuations at 10% CO_2 is depicted, with a standard deviation of 0.079%, in the right part of Fig. 4. Therefore, the sensor is qualified with a good measurement stability, with respect to the full scale range of 20%.

To validate the suitability of the sensor, the experiments was performed on two types of bacteria, *Bacillus subtilis (B. subtilis)* and *Stapbylococcus aureus (S. aureus)*, both being of importance in food and drug safety, in sealed vials and fostered in culture-medium based on beef extract and tryptone soya agar. The sensor was put in a temperature-controlled incubator. The experiments were conducted at ~35 °C and

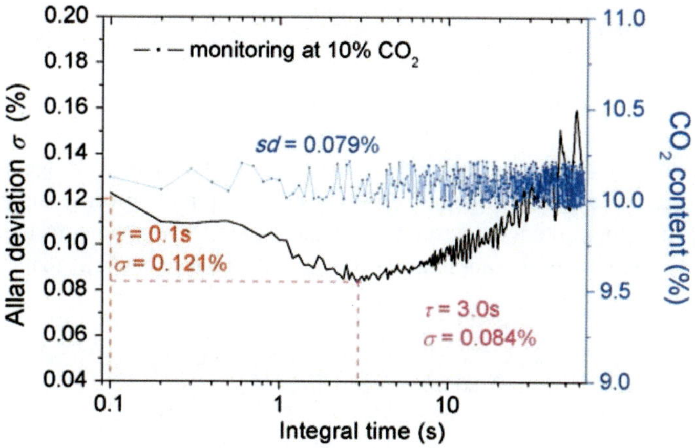

Fig. 3 Allan deviation and fluctuations at the concentration of 10%

Fig. 4 Microbial growth curves of *B. subtilis* and *S. aureus*

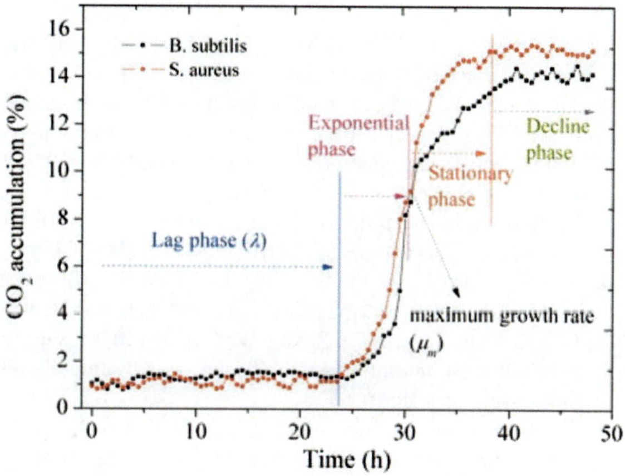

~65% RH. Two sealed vials were mounted on the rotating stage and were rotated every half an hour.

The experiment was conducted for about seven days, and the record data of the former two days was plotted in Fig. 4. At the beginning of measurement, the content of CO_2 in the headspaces of the vials was ~1.0%, much higher than that in air. Regardless of the difference of the vials and bacteria, the CO_2 contents in the headspaces of the samples change in a similar way and keep almost unchanged in the first 24 h of monitoring. This period is called lag phase. After that, CO_2 began to accumulate and reached the maximum at ~38 h. The maximal growth rate of CO_2 contents occurred after 30 h. It represents the same growth trend for the microbial organisms and shows the lag time of *S. aureus* is ~24 h, a little bit shorter than that of *B. subtilis*, and they reach decline phase at ~38 h, which almost agrees with the growth laws of microbiology.

4 Conclusion

We have demonstrated a TDLAS sensor operating at 2004.1 nm, combined with WMS technology, for monitoring the CO_2 concentration inside the microbial culture vials. The sensor owns a minimum detection limit of ~0.084% in the range of 0–20%. The experimental results show the sensor is suitable for monitoring the microbial growth by the CO_2 content detection in the headspace. It's very stale, fast and noninvasive, thus has a potential application in food and pharmaceutical manufacturing areas.

Acknowledgements This work was supported by National Natural Science Foundation of China under grant Nos. 41605101 and 41805108.

References

1. Danilović B, Savić D, Cocola L, Fedel M, Poletto L (2018) Determination of CO_2 content in the headspace of spoiled yogurt packages. J Food Qual 8121606
2. Moriaux A, Vallon R, Parvitte B, Zeninari V, Liger-Belair G, Cilindre C (2018) Monitoring gas-phase CO_2 in the headspace of champagne glasses through combined diode laser spectrometry and micro-gas chromatography analysis. Food Chem 264:255–262
3. Hu P, Pan S, Chen Z, Guo G, Zhang J, Li G, Chen N (2017) Hydrogen peroxide detection based on near-infrared TDLAS and Herriott cell. In: 16th international conference on optical communications and networks (ICOCN). Wuzhen, China, pp 1–3
4. Yang H, Chen J, Luo X, Liu C, Qi D, Xin X (2019) Leakage detection of closed vials based on two-line water-vapor TDLAS. Measurement 135:413–417
5. Cai T, Wang G, Cao Z, Zhang W, Gao X (2014) Sensor for headspace pressure and H_2O concentration measurements in closed vials by tunable diode laser absorption spectroscopy. Opt Lasers Eng 58:48–53
6. Cocola L, Fedel M, Giuseppe Tondello G, Frazzi G, Poletto L, Bardenshtein A (2018) Design and evaluation of an in-line system for gas sensing in flow-packed products. Food Packag Shelf Life 17:91–98
7. Brueckner D, Roesti D, Zuber UG, Schmidt R, Kraehenbuehl S, Bonkat G (2016) Comparison of tunable diode laser absorption spectroscopy and isothermal micro-calorimetry for non-invasive detection of microbial growth in media fills. Sci Rep 6(1):27894
8. Peng Z, Ding Y, Chen L, Li X, Zheng K (2011) Calibration-free wavelength modulated tdlas under high absorbance conditions. Opt Express 19(23):23104
9. Cocola L, Allermann H, Fedel M, Sonderby S, Tondello G, Bardenstein A (2016) Validation of an in-line non-destructive headspace oxygen sensor. Food Packag Shelf Life 9:38–44
10. Zarin AS, Chakraborty AL (2015) Absolute concentration measurements of bacterial CO_2 emission using a 2004 nm vertical cavity surface emitting tunable diode laser. Workshop on recent advances in photonics (WRAP), Bangalore, India

Enhanced Performance of Carbon Quantum Dots Based Organic Solar Cells

Wenhao Fan, Hairong Li, Huan Zhang, Lina Cai, Jianan Wang, Xudong Wang, Yongchang Wang, Ying Tang, and Yuzhe Song

Abstract Organic polymer solar cells (OSCs) have become research hotspot due to its low price, flexible preparation, solution preparation, translucency and so on in recently years. Adding quantum dots to an organic solar active layer to prepare a ternary structure becomes a promising approach. Here, we embedded carbon quantum dots into active layers to ameliorate the PCE of organic solar cells. The results of electron microscopy show that the CQDS are spherical and the average particle size is below 10 nm. Absorption spectra (ABS) of pure film layer and CQDs added film demonstrates that the active layer embedded with CQDs has better absorption strength than the active layer without CQDs added. The FF of carbon quantum dot-added device is 34%, the J_{SC} is 15.2 mA/cm^2, the V_{OC} is 0.55 V, and the PCE reached 2.81%.

Keywords Organic solar cells · Carbon quantum dots · Ternary active layer

W. Fan · H. Li (✉) · H. Zhang · L. Cai · J. Wang · X. Wang · Y. Wang
School of Physical Science and Technology, Lanzhou University, South Tianshui Road 222#, Lanzhou 730000, China
e-mail: lzulihairong@163.com

H. Li
Key Laboratory of Special Function Materials and Structure Design, Ministry of Education, Lanzhou University, South Tianshui Road 222#, Lanzhou 730000, China

Y. Song
Institute of Sensor Technology, Gansu Academy of Sciences, Gansu, China

Y. Tang
College of Optical and Electronic Technology, China Jiliang University, Hangzhou 310018, P.R. China

Y. Peng and X. Dong (eds.), *Proceedings of 2019 International Conference on Optoelectronics and Measurement*, Lecture Notes in Electrical Engineering 726,
https://doi.org/10.1007/978-981-33-4110-4_23

1 Introduction

Organic solar cells have become research hotspot due to its low price, low toxicity, low processing temperature, translucency and light mass for manufacturing on lightweight pliable substrates [1]. By designing new materials and device structures, the performance of OSCs has been greatly improved. To date, the PCE of small molecules or polymers OSCs has been reached to 11% [2], especially for non-fullerene OSCs, the power conversion efficiency (PCE) exceeds 14% [3]. However, for the commercialization of organic photovoltaics in the future, PCE needs to be further improved. Improving short-circuit current density (JSC) could play a part to enhance PCE. For example, owing to a vast majority conjugated polymer materials have many disadvantages, for example, the exciton diffusion length is short, and carrier mobility was limited, which severely hinders the thickness of the active layer in devices, turning into a bottleneck for further development of organic solar cells [4, 5]. Therefore, it is necessary to find a way to improve absorptivity in the active coating [6]. Optical nanocrystals [7], dielectric particles [8] and metal nanoparticles (NPs) [9] have been published to improve light absorption in OSCs.

The tandem OSCs have two or more active layers with different absorption window and two separate OSCs connected in parallel, which structure can widen the absorption broadband of the P3HT: PC61BM film [10, 11]. However, the preparation of serial and parallel OSCs requires a very complicated process, and the high cost of such devices had also become an important reason for limiting their application. These problems can be solved by preparing ternary OSCs. This method is achieved by adding extra light-absorbing materials to the photoactive layer with the acceptor and the donor. While achieving a wide absorption spectrum, it does not need to more complex experimental processes. The ternary structure provides an interesting way to ameliorate the PCE of cells. Ternary OSCs has caused a significant increase in Jsc, Voc and FF, leading to a significant improvement in performance [12].

In this paper, we embedded CQDs into organic solar cells in order to prepare OSCs with P3HT:PC_{61}BM: CQDs active layer structure. The morphology and crystal lattice of CQDs were characterized by transmission electron microscopy (TEM), and optical absorption spectra (ABS) of the films were tested by Shimadzu UV-3600 UV/VIS/NIR spectrophotometer, and the short circuit current density (Jsc), open circuit voltage (VOC), fill factor (FF), and PCE of devices were tested and discussed. The PCE of the optimized device is 2.81%, presenting 33% higher than the 2.12% of the reference device.

2 Experimental

2.1 Preparation of CQDs

The carbon quantum dots were composed by microwave chemical reaction [13–15]. Add urea and citric acid with the ratio of 2:1, put it into 20 mL of deionized water and stirred the solution till it turns to transparent. Then put the transparent mixture into a microwave oven to get the dark charred material. Subsequently, the reaction product was annealed for 2 h. At last, add appropriate amount of ethanol into the black solid to obtain the CQDs solution.

2.2 Devices Fabrication

The structure of glass/ITO/PEDOT: PSS/P3HT:PC_{61}BM: CQDs/Al and glass/ITO/PEDOT: PSS/P3HT: PC_{61}BM/Al devices were prepared, respectively. The device is prepared on an etched ITO glass sheet. Use ethanol, acetone and other detergents to clean the substrate, after that, the ITO glass was dried with a vacuum drying oven. The substrates were irradiated with UV light for 15 min before use to remove bacteria on the ITO surface and enhance the adhesion of PEDOT: PSS on the substrate. The PEDOT: PSS film was spin coated at 4000 r for 30 s, and annealed in air for a few minutes. Next, the rotation speed was set to 1500 rpm and the spin coating time was 40 s, and the prepared P3HT: PC_{61}BM and P3HT: PC_{61}BM: CQDs film were spin-coated on the annealed PEDOT: PSS film, and then annealed at 100 °C for 15 min. Ultimately, 80 nm Al was thermally deposited as the cathode of the device under the vacuum level of 10^{-4} Pa.

3 Results and Discussion

Figure 1 shows the Two-dimensional structure diagram of the device. From the bottom, it is obviously that the glass/ITO substrate, the PEDOT: PSS film, the P3HT: PC_{61}BM: CQDs layer, and the metal oxide as the cathode by thermal evaporation. The structure of pure devices is ITO/PEDOT: PSS/P3HT: PC_{61}BM/Al, and the optimized devices structure is ITO/PEDOT: PSS/P3HT: PC_{61}BM: CQDs/Al.

The TEM image of CQDs can be found in Fig. 2, the photograph of CQDs exhibiting that the shape of CQDs is granular. The diameter of CQDs are approximately 5–7 nm, and the lattice of CQDs is 0.15 nm.

The absorbance (ABS) of active layer film and CQDs optimized active layer film is shown in Fig. 3. The absorption spectrum was measured between 300 and 600 nm. It can be found that the CQDs embedded active layer shows better light absorption intensity than the active layer without CQDs, which indicates that the

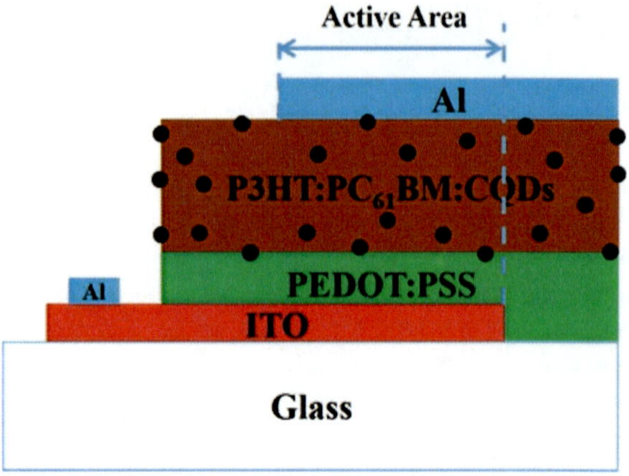

Fig. 1 Two-dimensional structure diagram of the device

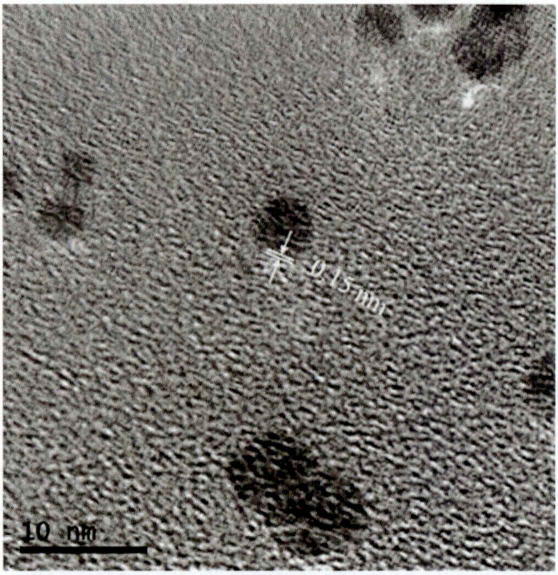

Fig. 2 The TEM picture of CQDs

embedding of CQDs promotes the light absorption of the active layer. Thus, the devices based on CQDs embedded film have better performance than the devices based on ITO/PEDOT: PSS/P3HT: $PC_{61}BM$ film.

Figure 4 shows the J-V curve of the polymer cells with increasing amounts of CQDs (from 0 to 2 mg/mL) in the active layer under AM 1.5 G illumination. Figure 4 demonstrate that the short circuit current density and open circuit voltage of CQDs added devices were better than the reference devices. The J_{SC} of device without CQDs is 13.5 mA/cm^2, and the open circuit voltage is 0.49 V, while the short circuit current density of device with CQDs is 15.2 mA/cm^2, the V_{OC} of device with CQDs is 0.55 V. Whereas, after continuing to add the concentration of CQDs in the active

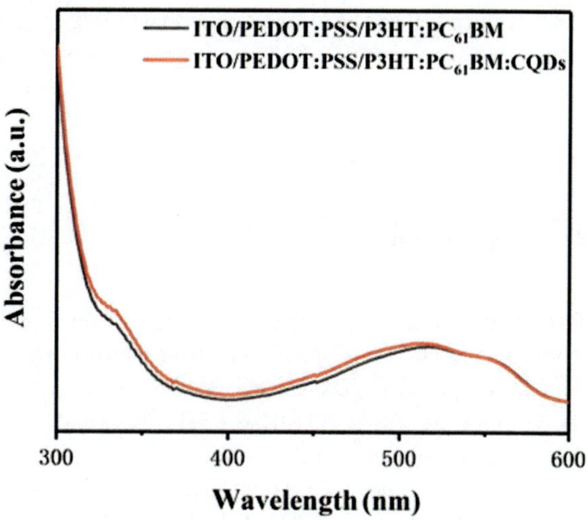

Fig. 3 UV–vis absorption spectra of ITO/PEDOT: PSS/P3HT: $PC_{61}BM$ film and ITO/PEDOT: PSS/P3HT: $PC_{61}BM$: CQDs film

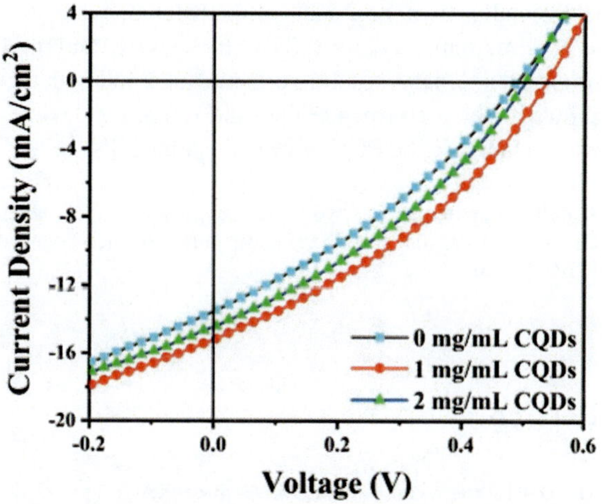

Fig. 4 J-V characteristics of CQDs with different CQDs ratios in the active layer under AM 1.5 G illumination

layer, the device performance began to decline, which may be related to the doping of CQDs in the active layer.

Table 1 shows the device performance parameters of CQDs doped at various

Table 1 Device parameters with different CQDs ratios in the active layer under AM 1.5 G illumination

CQDs content (mg/mL)	J_{SC} (mA/cm^2)	V_{OC} (V)	FF (%)	Rs (Ω)	Rsh (Ω)	PCE (%)
0	13.5	0.49	33	200	59.31	2.12
1	15.2	0.55	34	66.67	64.72	2.81
2	14.4	0.51	34	100	62.64	2.44

concentrations. It can be clearly found that the CQDs added device has a photoelectric conversion efficiency of 2.81%, which is 33% higher than the 2.12% of the reference device. This enhancement is primarily owing to the increment of current density caused by CQDs.

4 Conclusion

The organic solar cells were successfully prepared in the atmosphere and room temperature. The transmission electron microscopy (TEM) characterization shows that the prepared CQDs particles with a size of 5–7 nm. The absorption (ABS) shows that the ITO/PEDOT: PSS/P3HT: $PC_{61}BM$: CQDs ternary active layer film is enhanced between 300 and 520 nm. In summary, we prepared an organic solar cell with a ternary structure P3HT: $PC_{61}BM$: CQDs by adding a third component CQDs to the active layer. This ternary structure exhibits better performance than original devices based on P3HT: $PC_{61}BM$. The PCE of optimized device is 2.81%, which is 33% taller than the 2.12% of the reference device. In optimized devices, The $J_{SC} = 15.2$ mA/cm^2, $V_{OC} = 0.55$ V, FF = 34%, and the PCE = 2.81%. Therefore, we proposed a ternary structure organic solar cell with CQDs doped. This new structure shows better performance than the reference device. This method provides a new way to increase the PCE of organic solar cells.

Acknowledgements This work was supported by the National Natural Science Foundation of China (No. 6197405750272026) and Natural Science Foundation of Gansu Province (Grant No 17JR5RA180).

References

1. Dou L, You J, Hong Z, Xu Z, Li G, Street RA, Yang Y (2013) 25th anniversary article: a decade of organic/polymeric photovoltaic research. Adv Mater 25:6642–6671
2. Li M, Gao K, Wan X et al (2016) Solution-processed organic tandem solar cells with power conversion efficiencies >12%. Nat Photon 11:85–90
3. Xiao Z, Jia X, Ding L (2017) Ternary organic solar cells offer 14% power conversion efficiency. Sci Bulletin 62:1562–1564
4. Shrotriya V, Wu HE, Li G, Yao Y, Yang Y (2006) Efficient light harvesting in multiple-device stacked structure for polymer solar cells. Appl Phys Lett 88
5. Notarianni M, Vernon K, Chou A, Aljada M, Liu J, Motta N (2014) Plasmonic effect of gold nanoparticles in organic solar cells. Sol Energy 106:23–37
6. Raut HK, Ganesh VA, Nair AS, Ramakrishna S (2011) Anti-reflective coatings: a critical, in-depth review. Energy Environ Sci Home 4:3779–3804
7. Gomard G, Meng X, Drouard E, Hajjam KE, Gerelli E, Peretti R, Fave A, Orobtchouk R, Lemiti M, Seassal C (2012) Light harvesting by planar photonic crystals in solar cells: the case of amorphous silicon. J Opt 14

8. Shao P, Chen X, Guo X, Zhang W, Chang F, Liu Q, Chen Q, Li J, Li Y, He D (2017) Facile embedding of SiO_2 nanoparticles in organic solar cells for performance improvement. Org Electron 50:77–81
9. Xie F, Choy WCH, Wang CCD, Sha WEI, Fung DDS (2011) Improving the efficiency of polymer solar cells by incorporating gold nanoparticles into all polymer layers. Appl Phys Lett 99
10. Ameri T, Dennler G, Lungenschmied C, Brabec CJ (2009) Organic tandem solar cells: a review. Energy Environ Sci Home 2(347–363)
11. Zuo L, Yu J, Shi X, Lin F, Tang W, Jen AK (2017) High-efficiency nonfullerene organic solar cells with a parallel tandem configuration. Adv Mater 29
12. Bi P, Hao X (2018) Versatile ternary approach for novel organic solar cells: a review. Solar RRL 3
13. Sun M, Qu S, Hao Z, Ji W, Jing P, Zhang H, Zhang L, Zhao J, Shen D (2014) Towards efficient solid-state photoluminescence based on carbon-nanodots and starch composites. Nanoscale 6:13076–13081
14. Zhu H, Wang X, Li Y, Wang Z, Yang F, Yang X (2009) Microwave synthesis of fluorescent carbon nanoparticles with electrochemiluminescence properties. Chem Commun 5118–5120
15. Zhai X, Zhang P, Liu C, Bai T, Li W, Dai L, Liu W (2012) Highly luminescent carbon nanodots by microwave-assisted pyrolysis. Chem Commun 48, 7955–7957

High Performance Gas Sensor Based on ZnO/CuO Heterostructures

Lina Cai, Hairong Li, Huan Zhang, Wenhao Fan, Jianan Wang, Yongchang Wang, Xudong Wang, Ying Tang, and Yuzhe Song

Abstract The ZnO/CuO heterostructures were synthesized by two simple methods. The structural and morphological characterizations of the as-deposited samples were well evaluated by using X-ray diffraction (XRD) and field emission scanning electron microscopy (FESEM). The gas sensing characteristics of the pure CuO and ZnO/CuO heterostructure were discussed. By comparison, the ZnO/CuO heterostructure sensor synthesized by water bath deposition has a higher response than pure CuO obviously. At the optimal working temperature, the response of the CuO-ZnO gas sensor to ethanol (100 ppm) reached 31.2, which was about 4.3 times higher than the original CuO nanoparticles. Besides, this heterojunction also showed good selectivity towards ethanol. The ZnO/CuO heterostructure provides a feasible method for preparing high performance ethanol sensor

Keywords Gas sensor · Heterostructure · CuO · ZnO · Ethanol

L. Cai · H. Li (✉) · H. Zhang · W. Fan · J. Wang · Y. Wang · X. Wang
School of Physical Science and Technology, Lanzhou University, South Tianshui Road 222#, Lanzhou 730000, China
e-mail: lzulihairong@163.com

H. Li
Key Laboratory of Special Function Materials and Structure Design, Ministry of Education, Lanzhou University, South Tianshui Road 222#, Lanzhou 730000, China

Y. Song
Institute of Sensor Technology, Academy of Sciences, Gansu, China

Y. Tang
College of Optical and Electronic Technology, China Jiliang University, Hangzhou 310018, P.R. China

Y. Peng and X. Dong (eds.), *Proceedings of 2019 International Conference on Optoelectronics and Measurement*, Lecture Notes in Electrical Engineering 726,
https://doi.org/10.1007/978-981-33-4110-4_24

1 Introduction

Effective monitoring and detection of harmful gases from the atmosphere is essential to human health [1]. Due to the advantages of low cost and simple manufacturing, gas sensors of metal oxide semiconductors have been widely used in the monitoring and detection of toxic and harmful gases [2–4]. The CuO with a narrow band gap (Eg = 1.2 eV) is extensively explored for their gas sensing behavior [5]. In order to enhance the performance of the CuO gas sensor, many researches were carried out [6]. Recently, a lot of research has been focused on the design of the high sensitivity gas sensors based on the use of more than one material to form heterostructures. Researchers are continuing to demonstrate that the combination of different materials, the use of different synthetic methods or the design of different nanostructures affects the gas sensing performance of the system [7]. Therefore, constructing two-layer heterostructure with ZnO and CuO might be useful way to prepare high-response gas sensor.

In this paper, two-layer ZnO/CuO heterostructures were synthesized by using different methods. Gas sensing performance of the pure CuO and ZnO/CuO heterostructure sensors were investigated. The results indicate that ZnO/CuO heterostructure exhibit excellent sensing properties towards ethanol with improved response and selectivity.

2 Experimental

2.1 Preparation of CuO and ZnO/CuO Heterostructures

The Zn thin film was grown by using electrodeposition at room temperature. The classic three electrodes are used in electrochemical experiments. First, $ZnSO_4 \cdot 7H_2O$ was dissolved in 150 mL of DW at room temperature as an electrolytic solution. The working electrode is the pre-cleaned ITO substrate, the reference electrode is KCl, and the auxiliary electrode is graphite paper. Set the appropriate polarization current and the deposition time. After completion, the substrate was rinsed with DW and was thermally oxidized in air for 50 min at 350 °C to form ZnO thin film.

An simple hydrothermal method was used to prepare ZnO nanorods, as previously reported [8]. First, zinc acetate ethanol solution was spin-coated on the pre-cleaned glass substrate and then annealed at 200 °C for 10 min. The reaction solution is a certain amount of $Zn(CH_3COO)2 \cdot 2H_2O$ and C6H12N4 aqueous solution. Thereafter, the substrate was placed upward in the above solution at 90 °C for 1 h. The sample was dried at 60 °C for 8 h.

Then, the CuO material was synthesized using the same hydrothermal process on the glass substrate, the ITO substrate grown with ZnO thin film and the glass substrate grown with ZnO nanorods. The as-prepared samples were labeled as 1#, 2# and 3#, respectively.

2.2 *Characterization*

X-ray diffraction (XRD) of as-synthesized nanostructures samples were investigated on Rigaku D/max-2400 with Cu Kα radiation. The surface morphology and elemental composition of the as-deposited material was obtained using field emission scanning electron microscopy (FE-SEM, Apreo S). The sensing characteristics of the gas sensor were measured using the QJC-II series Intelligent Test Meter. The schematic diagram of gas sensor test system has been reported in our previous work [9].

2.3 *Fabrication and Measurement*

To fabricate the CuO and the ZnO/CuO sensors, the as-deposited material was thoroughly dissolved in ethanol to obtain a uniform slurry. Then, the slurry was coated on alumina tube. The structure diagram of the gas sensor has been listed in the previously published work. Gas sensing measurements were carried out at about 30–40% RH in atmospheric air. The operating temperature of the gas sensor was controlled from 116–151 °C. The response (R) of the sensor was defined as follows, R = Rg/Ra, where Ra was the resistance in the target gas and Rg was the resistance in air.

3 Results and Discussion

3.1 *Structural and Morphological Characteristics*

XRD patterns of the pure CuO and the ZnO/CuO heterostructure prepared by water bath are shown in Fig. 1. The black curve displays XRD pattern of the pure CuO, it contains two kinds of peaks. The planes (0 0 2), (1 1 1), (−2 0 2), (2 0 2), (−1 1 3), (0 2 2) and (1 1 3) which were corresponded to standard PDF#45-0937 of the CuO crystal. The (1 1 0), (1 2 1) planes (marked in olive) of SiO2 was given by standard PDF#45-1374. The peak of SiO2 may come from the used glass substrate. The red curve displays XRD pattern of ZnO/CuO heterostructure prepared by water bath. The peaks corresponded to (1 0 0), (0 0 2), (1 0 1), (1 0 2), (1 1 0), (1 0 3), (2 0 0), (1 1 2), (2 0 1) and (2 0 2) planes of the ZnO crystal were in good agreement with standard PDF#36-1451. The existence of diffraction peaks corresponding to CuO and ZnO observed in the XRD spectrum indicates the synthesis of ZnO/CuO composite heterostructure.

Figure 2 shows the morphology and microstructure of the as-synthesized samples by FESEM image. It can be seen from Fig. 2a that the prepared pure CuO material is composed of nano-particle clusters. Figure 2b shows the morphology of the ZnO layer in the ZnO/CuO heterojunction prepared by a water bath method, as can be seen that the ZnO layer consisting of stacked ZnO nanorods. The surface morphology of

Fig. 1 XRD patterns of pure CuO and ZnO/CuO heterostructure

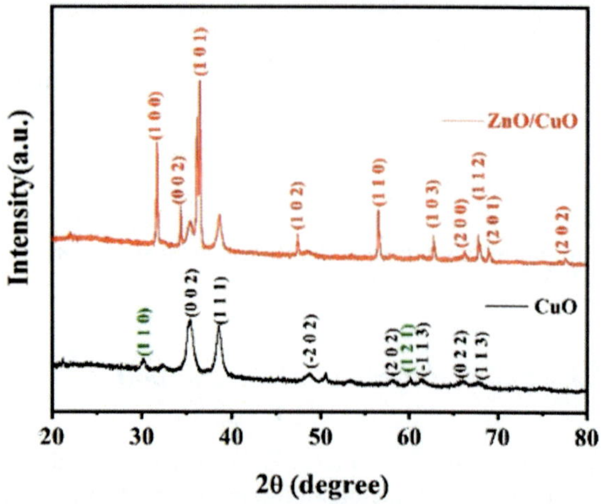

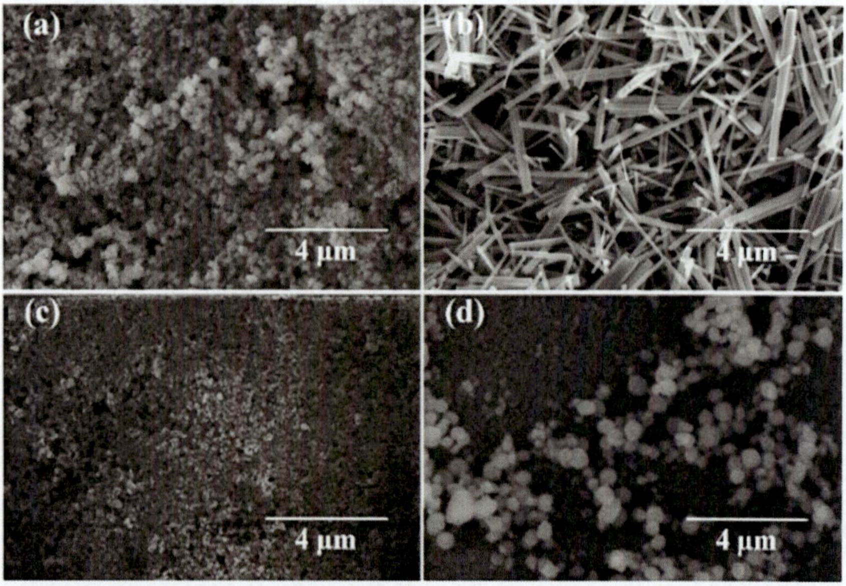

Fig. 2 FESEM images of **a** the pure CuO, **b** the ZnO, **c** the ZnO/CuO heterostructure prepared by water bath and **d** the ZnO/CuO heterostructure prepared by electrodeposition and water bath

the ZnO/CuO heterojunction prepared by the water bath method is shown in Fig. 2c, which is also composed of nanoparticle clusters. However, the size of the ZnO/CuO nanoparticles is greatly reduced compared to pure CuO, which significantly increases the gas sensitivity of the material. Figure 2d shows that the ZnO/CuO nanostructure

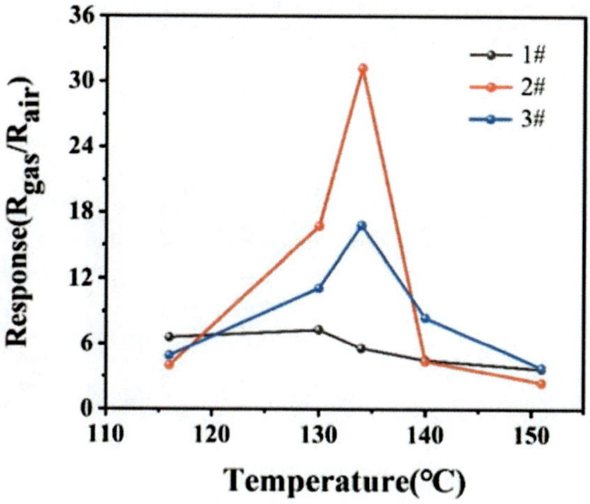

Fig. 3 Response of the 1#, 2#, and 3# sensor from 116 to 158 °C toward 100 ppm ethanol

prepared by the two-step process consists of copper oxide nanospheres and a zinc oxide film.

3.2 *Gas Sensing Characteristics*

Before testing, the 1#, 2#, and 3# sensors were preheated for 30 min. In order to obtain the optimal working temperature, the response of the 1#, 2#, and 3# sensors to 100 ppm ethanol gas were investigated at different temperatures. Figure 3 shows the response of the three sensors to 100 ppm ethanol at the corresponding working temperature. The pure CuO gas sensor gets the relatively low operating temperature of 130 °C. The response of ZnO/CuO heterostructures sensors reaches its maximum at the temperature of 134 °C. Clearly, the maximum response value of 2# ZnO/CuO based sensor to ethanol reached 31.2 at 134 °C, which is about 4.3 times higher than that of pure CuO nanoparticles (R = 7.3). According to the observation results, the ZnO/CuO heterostructures sensors displayed an enhanced response to ethanol.

Subsequently, the gas response of the 1#, 2# and 3# sensors towards 100 ppm target VOCs gas are shown in Fig. 4. It is obvious that all the sensors exhibit the highest response to ethanol and a relatively lower response to other test gases. The sensors based on ZnO/CuO heterostructures have improved ethanol performance compared to pure CuO, which is related to the formation of heterostructures. Furthermore, the response of sensor based on the 2# ZnO/CuO to 100 ppm ethanol gas is higher than the other sensors tested. The results indicated that the 2# ZnO/CuO sensor show a good selectivity for ethanol.

Figure 5a-c displays the dynamic resistance curves of the 1#, 2# and 3# gas sensors to different concentrations of ethanol under the optimum temperature. With the concentration of ethanol gas increasing, the response of all sensors increases,

Fig. 4 Response of the 1#, 2#, and 3# sensor to 100 ppm ethanol, acetone, formaldehyde, methanol and xylene gases

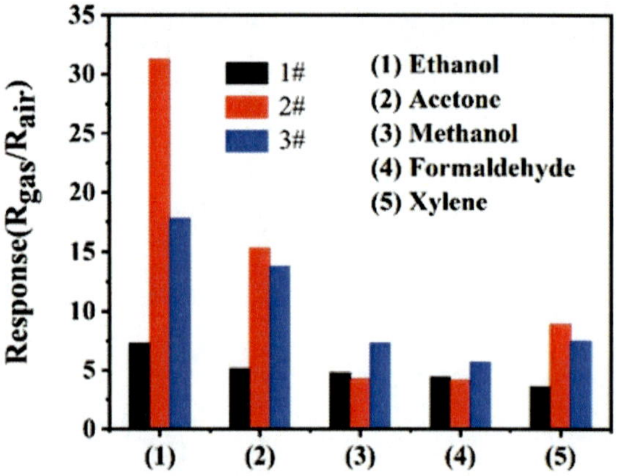

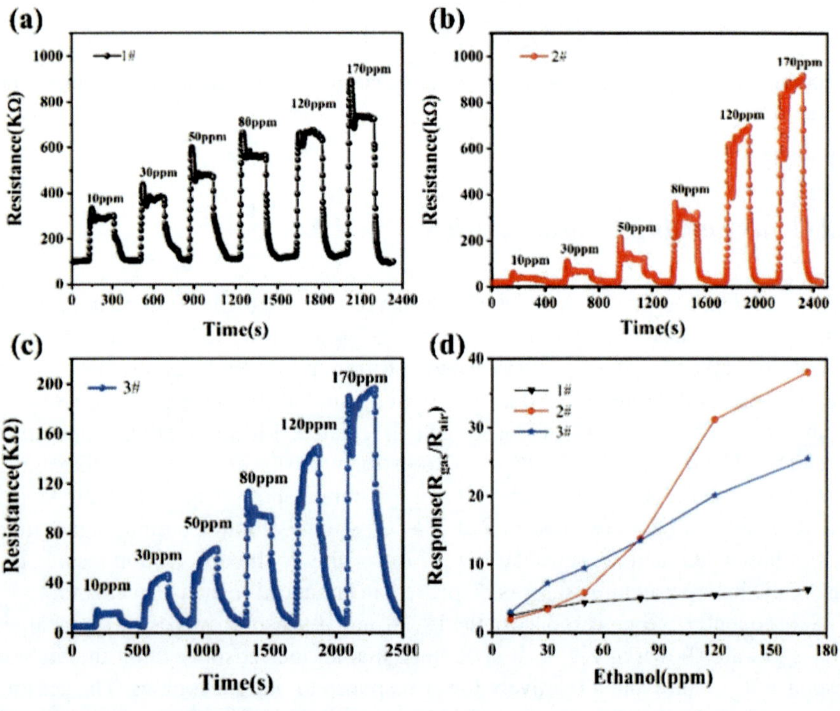

Fig. 5 Dynamic response curves of **a** the 1#, **b** the 2#, and **c** the 3# sensor to different concentrations of ethanol, **d** Response of the 1#, 2#, and 3# sensor to different concentrations of ethanol

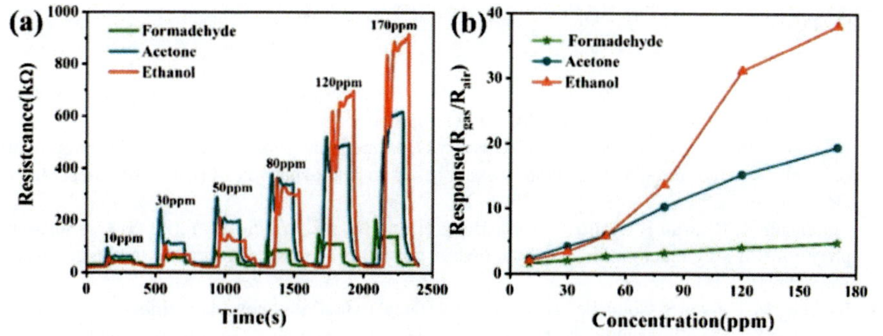

Fig. 6 **a** Dynamic response curves of the 2# sensor towards ethanol, acetone and formaldehyde. **b** Response of the 2# sensor towards ethanol, acetone and formaldehyde at different concentration

which is apparently presented in Fig. 5d. And the response of 2# ZnO/CuO sensor is significantly higher than other sensors, when the concentration of ethanol is over 80 ppm.

Figure 6a shows that the dynamic response and recovery characteristics of the 2# sensor under different concentration of ethanol, acetone and formaldehyde from 10 to 170 ppm at 134 °C. Figure 6b shows the corresponding response value of the 2# sensor under different concentration of ethanol, acetone and formaldehyde from 10 to 170 ppm at 134 °C. The test results show that the prepared device has good dynamic stability to these three gases.

4 Conclusion

In short, the ZnO/CuO heterostructures were synthesized using two simple methods. Compared with pure CuO nanoparticles, the 2# ZnO/CuO heterostructure sensor exhibits significantly improvement in the ethanol sensing response. Upon exposure to 100 ppm of ethanol gas at 134 °C, the response value of 2# ZnO/CuO heterostructure sensor is up to 31.2, which is about 4.3 times higher than the pure CuO nanoparticles. Besides, the response of ZnO/CuO heterojunction to ethanol is much higher than other measured gases. The ZnO/CuO heterojunction shows good selectivity towards ethanol. The obtained gas sensing performance demonstrates that gas sensor based on the ZnO/CuO heterostructure prepared by water bath method can be a prospective candidate for ethanol detection in environmental monitor.

Acknowledgements This work was supported by the National Natural Science Foundation of China (No. 6197405750272026) and Natural Science Foundation of Gansu Province (Grant No 17JR5RA180).

References

1. Subha PP, Jayaraj MK (2019) Enhanced room temperature gas sensing properties of low temperature solution processed ZnO/CuO heterojunction. BMC Chem 13
2. Ahmad U, Jong-Heun L, Rajesh K, Al-Dossary O, Ahmed AI, Baskoutas S (2016) Development of highly sensitive and selective ethanol sensor based on lance-shaped CuO nanostructures. Mater Des 105:16–24
3. Prabhakar R, Rizwan K, Rafiq A, Yoon-Bong H, In-Hwan L, Yeon-Tae Y (2013) Gas sensing properties of single crystalline ZnO nanowires grown by thermal evaporation technique. Curr Appl Phys 13:1769–1773
4. Xicheng M, Haiyan S, Congsheng G (2013) Interfacial oxidation–dehydration induced formation of porous SnO2 hollow nanospheres and their gas sensing properties. Sensors Actuators B Chem 177:196–204
5. Jan D, Alexander Z, Ivan J, Mariana K, Kamil L (2015) Facile synthesis of CuO nanosheets via the controlled delamination of layered copper hydroxide acetate. J Colloid Interface Sci 452:174–179
6. Hyo-Joong K, Jong-Heun L (2014) Highly sensitive and selective gas sensors using p-type oxide semiconductors: overview. Sensors Actuators B Chem 192:607–627
7. Miller DR, Akbar SA, Morris PA (2014) Nanoscale metal oxide-based heterojunctions for gas sensing: a review. Sensors Actuators B Chem 204:250–272
8. Lori EG, Matt L, Joshua G, Franklin K, Justin CJ, Yanfeng Z, Saykally RJ, Peidong Y (2003) Low-temperature wafer-scale production of ZnO nanowire arrays. Angew Chem 42:3031–3034
9. Wang F, Li H, Yuan Z, Sun Y, Chang F, Deng H (2016) A highly sensitive gas sensor based on CuO nanoparticles synthetized via a sol-gel method. RSC Adv 6(83):79343–79349

Design of Stripe Projection System for 3D Measurement of Structured Light

Lei Zhang, Limin Hu, and Tengbiao Song

Abstract There are four common micro-electro-mechanical system (MEMS) micro-mirror driving methods: electrostatic driving, electromagnetic driving, electrothermal driving, and piezoelectric driving. Among them, the electrostatically driven MEMS micro-mirror has natural technical advantages in the field of 3D vision: it can achieve the smallest volume, the lowest power consumption, the process is relatively simple, the reliability and yield are high, and the cost is the lowest. A fringe projection system based on a MEMS micro-mirror is designed. After the laser beam passes through the MEMS micro-mirror at high speed, a fringe structured light pattern in the imaging area is generated. Select the peripheral SPI1 of the STM32F103C8T6 chip of the STM32 series of the single chip microcomputer micro control unit (MCU) as the host to exchange data with the slave where the MEMS micro-mirror is located, the program is written with software Keil uVision5 and driven by the MEMS. At the same time, the microcontroller outputs a pulse width modulation signal to drive the laser to generate a pulse width modulation waveform (PWM) waveform laser bundle. The experimental results show that the MEMS micro-mirror is controlled by a single-chip microcomputer for simple harmonic motion. The line laser pulse width can be controlled by the chip's timer. The optical path can be adjusted and the circuit can be driven to generate a laser pattern with red laser stripes and dark stripes interlaced in the imaging area. Has profound significance for 3D measurement technology.

Keywords Structured light · 3D measurement · MEMS · Laser · MCU

L. Zhang · L. Hu
Hangzhou HPWINNER OPTO Corporation, Hangzhou, Zhejiang, China

T. Song (✉)
College of Optical and Electronic Engineering, China Jiliang University, Hangzhou, Zhejiang, China
e-mail: Soonggg37@gmail.com

Y. Peng and X. Dong (eds.), *Proceedings of 2019 International Conference on Optoelectronics and Measurement*, Lecture Notes in Electrical Engineering 726,
https://doi.org/10.1007/978-981-33-4110-4_25

1 Introduction

With the development and demand of modern industrial production and science and technology, the shape of products has become increasingly complex, the production cycle has become shorter and shorter, and the accuracy of processing and assembly has become higher and higher, creating the need to obtain the three-dimensional shape of objects [1], Especially in the fields of reverse engineering, virtual reality, machine vision, biomedical monitoring, anthropometrics and many other fields [2–6], there are extremely extensive requirements for the measurement of three-dimensional shape of objects. The structured light measurement method originated in the 80s. By measuring the three-dimensional shape of the measured object by projecting structured light onto the measured object, the main idea is to observe the measured object from a direction different from the projection direction. And the triangular relationship between the measured object and the measured object to obtain the three-dimensional shape of the measured object [7].

Micro-electro-mechanical system (MEMS) micro-mirror is a typical optical microelectronic machine. The incident light beam can be reflected in a specific manner and time sequence, thereby realizing scanning imaging on the image plane. In addition, MEMS scanning micro-mirrors have also been widely used in laser confocal microscopy systems, optical switches, projection displays and other fields [8]. The initial technology choices depend largely on the application and performance requirements in terms of scan speed, scan angle, impact resistance, power consumption, and package compatibility [9]. Compared with traditional optical micro-mirrors, MEMS micro-mirrors have lower cost and easy mass production than MEMS devices, but also have better optical and mechanical performance, especially in terms of dynamic response and power consumption. As prominent [10].

A fringe structured light generation technology device based on MEMS was designed. As a fringe projection system structure for 3D measurement applications, a laser scanning micro-projection technology was selected to cooperate with a MEMS micromirror to generate ideal fringe structured light.

2 Principle

The one-dimensional MEMS micro-mirror projection system is mainly composed of a MEMS micro-mirror driving circuit, an optical module, and a laser driving circuit. Firstly driven by the MEMS, the MEMS micro-mirror is made to perform a simple harmonic motion at a certain angle under the control of a single-chip microcomputer; the point laser generated by the laser is diffused by the optical module into a line laser with a certain waveform. The width modulation can be generated by the chip's timer. The software chooses Keil uVision5 and then uses the FLYMCU programming program. After the high-speed vibration MEMS micro-mirror is irradiated, the laser beam will be reflected to the projection area and generated in the imaging area. Laser pattern of red laser stripes and dark stripes.

3 MEMS and MCU

MEMS laser scanning micro-projection technology is a cutting-edge technology that has only recently been developed. It has good applications in head-mounted displays, micro-projection display technology, handheld electronics, and automotive fields, and has good development space. There are three main components of a laser display system: a laser, a light modulator, and a beam scanner. The light emitted by the three primary color diodes passes through the filter lens and is combined into a full color, which can highly restore the color.

Compared with digital light processing (DLP), liquid crystal on silicon (LCOS), liquid crystal display (LCD) micro-projection technology, MEMS laser scanning micro-projection technology has a simpler structure and easier system integration. One-dimensional MEMS vibrating mirror projection system structure, as shown in Fig. 1.

MEMS laser scanning micro-projection technology has a high-power-primary-color laser light source, which is composed of full color. The color gamut we can see is theoretically more than 90%, which is much higher than other micro-projection display technologies. In addition, the laser has higher luminous efficiency, which means that when projecting the same brightness screen, it will consume less energy and its color reproduction is higher. Therefore, the laser scanning micro-projection technology is used in conjunction with the optical MEMS micro-mirror to complete the imaging effect.

Both the MEMS micro-mirror and the laser drive require a single-chip micro-computer. The STM32 series chips have better performance, lower cost and energy consumption, and faster computing speed. Therefore, the single chip microcomputer MCU model STM32F103C8T6 chip was selected. It has only 48 pins. The highest frequency is 72 MHz. Although it is small, it also has two peripheral SPIs and 32 general-purpose I/O ports. The minimum system schematic is shown in Fig. 2.

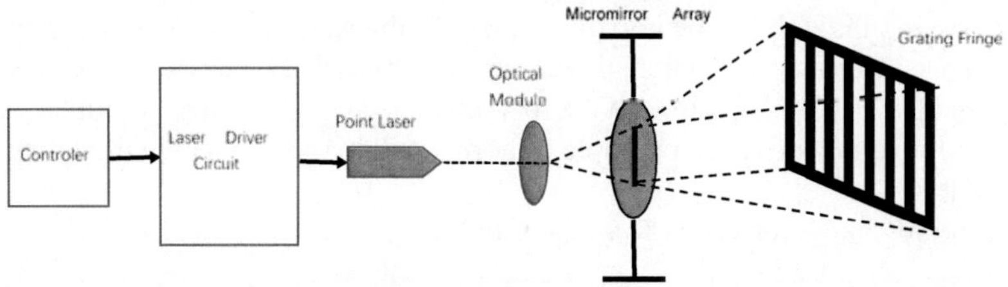

Fig. 1 Schematic diagram of the one-dimensional MEMS micro-mirror projection system

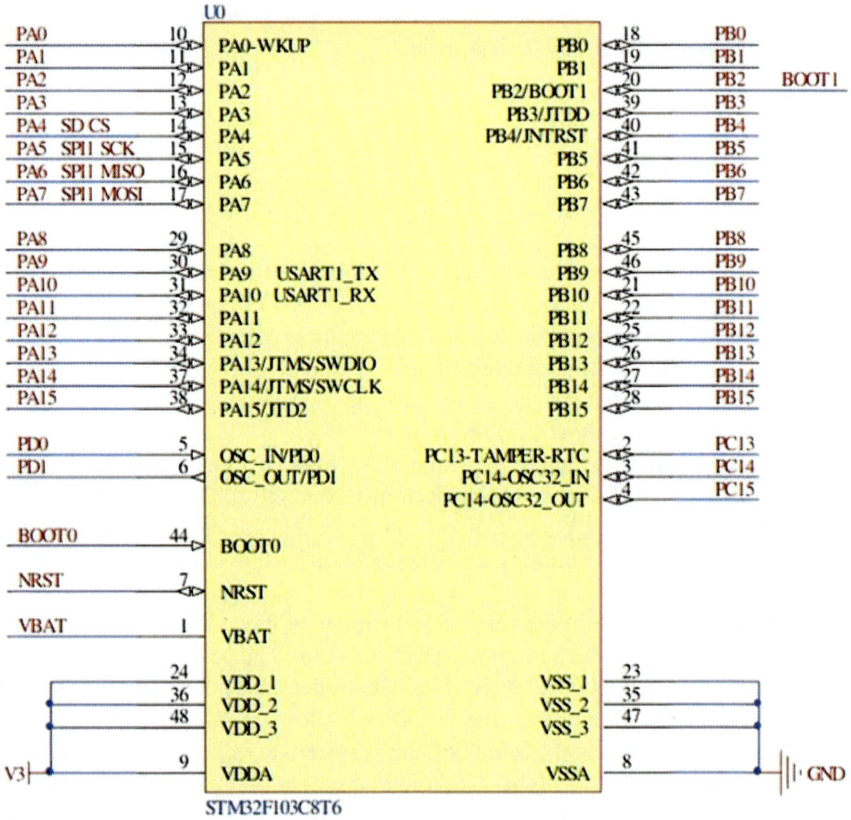

Fig. 2 STM32F103C8T6 chip minimum system structure principle diagram

3.1 *Laser Drive*

The laser is a red dot laser with transistor-transistor logic level (TTL) modulation signal. The spot size can be adjusted to facilitate imaging. Let the point laser emitted by the laser pass through a cylindrical optical lens. The vertical surface of the cylindrical mirror is the same as that of flat glass. The light does not diverge or converge when passing through it. The horizontal plane is the same as the imaging principle of the lens. After two refractions later, the cylindrical lens focuses the beam first, and then the beam diverges toward the forward cone area. A red line is formed when it hits the imaging area. The principle of horizontal imaging of a cylindrical optical lens is shown in Fig. 3.

According to the refraction formula of light:

$$n_1 \sin\theta_2 = n_2 \sin\theta_1 \tag{1}$$

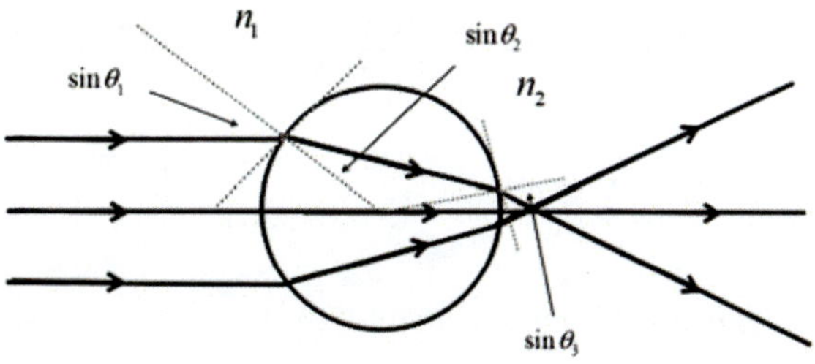

Fig. 3 Horizontal mirror map of cylindrical mirror

The refraction angle of the laser entering the lens is:

$$\sin\theta_2 = n_2\sin\theta_1/n_1 \tag{2}$$

According to the principle of an isosceles triangle, the second incident angle of the laser is the same as the first outgoing angle. The exit angle of the second refracted laser lens is:

$$\sin\theta_3 = n_1\sin\theta_2/n_2 \tag{3}$$

(3) After bringing into formula (2):

$$\sin\theta_3 = \sin\theta_1 \tag{4}$$

n_1 is the refractive index in the air, n_1 is the refractive index in the cylindrical lens, θ_1 is the incident angle θ_2 is the first exit angle, and $\sin\theta_3$ is the second exit angle.

Obviously, the final exit angle and incident angle of the laser are the same. Because the coherence of the laser is good, it can be considered as an ideal parallel light. According to the basic geometric principle, the laser beam passing through the center of the circle is used as the boundary. The parallel lights below the center of the circle are also emitted parallel to each other.

3.2 *Signal Read*

It is very convenient to send instructions to the MEMS module through the read and write functions of SPI1. When you need to configure the angle, you can send a 14-bit unsigned number to the mems module. You need to send the instruction 0XC00 when you query. The host will receive two 16-bit data. interval. The angle configuration is shown in Table 1:

The MEMS module outputs a zero-mark pulse at the beginning of a resonance cycle, and outputs an angle mark pulse each time a minimum angular interval is changed. Assume that the scanning angle output by the MEMS module at this time

Table 1 Send 14 unsigned numbers to P1130 through SPI1 read and write function

D15	D14	D13	D12	D11	D10	D9	D8	D7	D6	D5	D4	D3	D2	D1	D0
0	0	Output angle													
1	0	Angular interval													

is $\pm\theta$ and the angular interval is Δ. Let:

$$i = \theta/\Delta \tag{5}$$

Then in a simple harmonic cycle, that is, between receiving zero-mark pulses, a total of (4i + 1) angle mark pulses will be received. Then in a simple harmonic period, the angle θ_j corresponding to the j-th angle mark pulse sent by the MEMS module is:

$$(j - 1) * \Delta 1 < j \leq i + 1 \tag{6}$$

$$(2 * i - j + 1) * \Delta \mathrm{i} + 1 < j \leq 2i - 1 \tag{7}$$

$$0\mathrm{j} = 1, 2i + 1 \tag{8}$$

$$-(j - 2 * i - 1) * \Delta 2i + 1 < j \leq 3i + 1 \tag{9}$$

$$-(4 * i + 1 - j) * \Delta 3i + 1 < j \leq 4i + 1 \tag{10}$$

4 Experimental Part

The one-dimensional scanning module can calculate the scanning angle of the micromirror at each time point of the MEMS by sending the angle marking pulse and the zero-mark pulse. Mems micro-mirror stripe structured light production system is shown in Fig. 4.

4.1 MEMS and Microcontroller Selection

The MEMS device uses a one-dimensional laser scanning module P1130 developed by Xi'an ZHISENSCR. As shown in Fig. 5, it uses an electrostatically driven MEMS micro-mirror. The connection method of P1130 is to use the connector female seat,

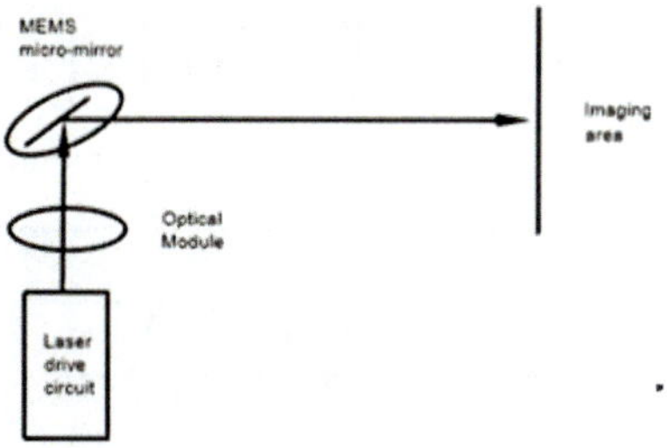

Fig. 4 MEMS micro vibrating mirror stripe structure light production system

Fig. 5 One-dimensional laser scanning module P1130 physical map developed by Xi'an Zhiwei sensor

and use the HRS connector model DF37NB-10DS-04V. As shown in Fig. 6, the specification is 10PIN and the PIN pitch is 0.4 mm.

When P1130 is connected to the microcontroller, P1130 is set as the slave device of the microcontroller, and the peripheral SPI (Serial peripheral interface) of the microcontroller controls P1130. First, ensure that they share the same ground to ensure normal communication between them. The SPI NSS pin of STM32F103C8T6 chip is connected to the SPI CS of P1130. When the chip select signal of the master is high and the slave is low, communication between the master and the slave can be performed. The SPI SCK pins of the two are connected to each other to synchronize

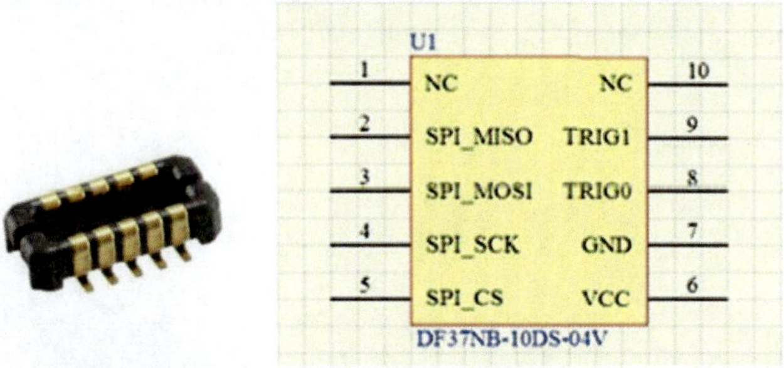

Fig. 6 The physical diagram and schematic diagram of the HRS connector model DF37NB-10DS-04V

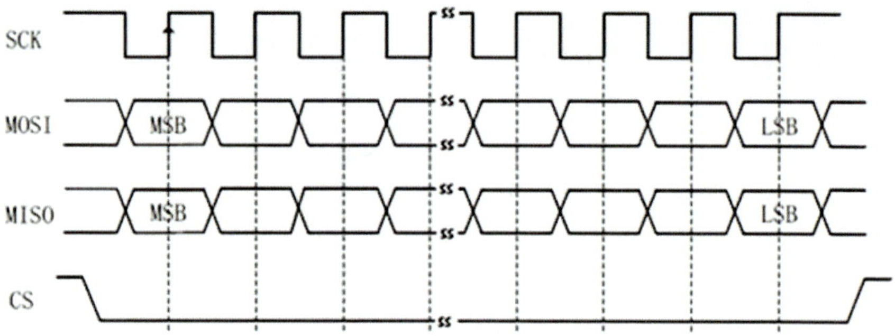

Fig. 7 SPI communication time sequence diagram

the clock signal of the microcontroller and P1130; the SPI MOSI of the microcontroller is connected to the SPI MISO of P1130. When it is working, the microcontroller outputs signal and P1130 receives; when the SPI 32 of STM32 single chip microcomputer is connected to the MOSI of P1130 the opposite is true. SPI communication timing diagram, as shown in Fig. 7:

4.2 Program Input

Because the STM32F103C8T6 chip minimal system itself does not have USB to TTL function, the chip's peripheral SPI cannot directly communicate with the personal computer, so an external FT232 module is needed to implement this function, as shown in Fig. 8. The BOOT0 of the chip is connected to a 3.3 V power supply, and the BOOT1 is grounded. The TXD interface of the FT232 module is connected to the RXD interface of the chip, and the RXD interface is connected to the TXD interface of the chip, and the GND interface is connected. The 3.3 V port of the FT232 module can provide power to the development board and restart the chip after the connection is completed. It will enter SPI mode. After that, the serial port SPI programming software FLYMCU is used, and the bit rate is set to 115,200. After searching for a free com port, the program can be downloaded.

Fig. 8 FT232 modular

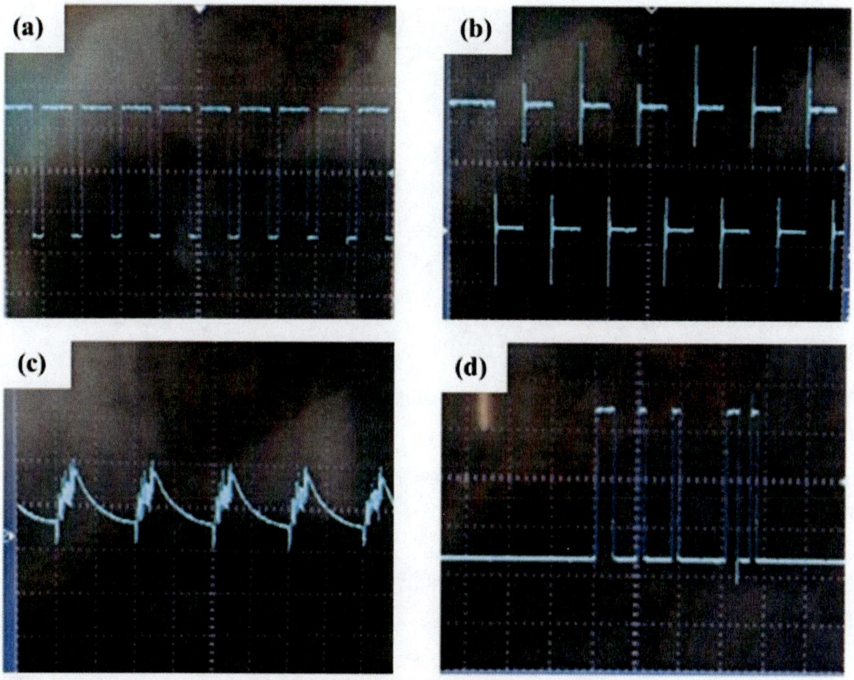

Fig. 9 When sending data, **a** PA4, **b** PA5, **c** PA6, **d** PA7 pin waveform

4.3 Signal Detection

When the peripheral SPI1 sends data to P1130, PA5 generates a clock signal, the PA4 pin first becomes a low voltage state, and the host SPI1 transmits the content to P1130. After the content transfer is completed, the chip select signal is set to high level. PA7 has a square wave signal, because P1130 does not send a response to the peripheral SPI1, so PA6 has no signal. In order to make the waveform of each pin easy to observe, put the program that sets the output angle and angle interval of the module into a while (1) loop, then add a 1 ms delay after each data read, and then burn Go to the chip to test the waveform. When the peripheral SPI1 receives data from P113, the PA6 pin also has waveform output. Figure 9 shows the waveforms of the PA4, PA5, PA6, and PA7 pins when sending data.

4.4 Generation of Striped Structured Light

The overall environment setup is shown in Fig. 10. After the driving of the MEMS and the laser are completed, the optical path is placed, and the spot laser beam emitted

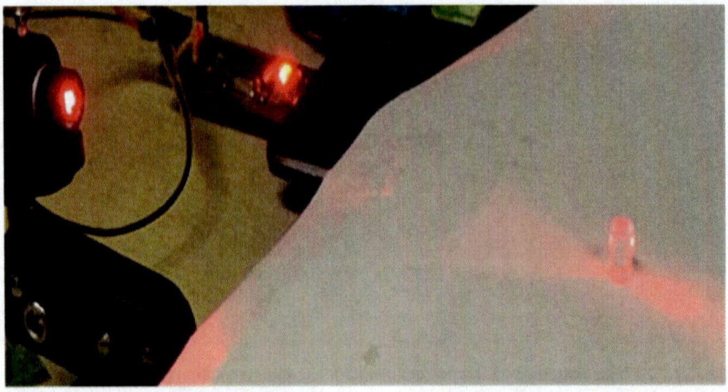

Fig. 10 Physical picture of the overall structure of the system

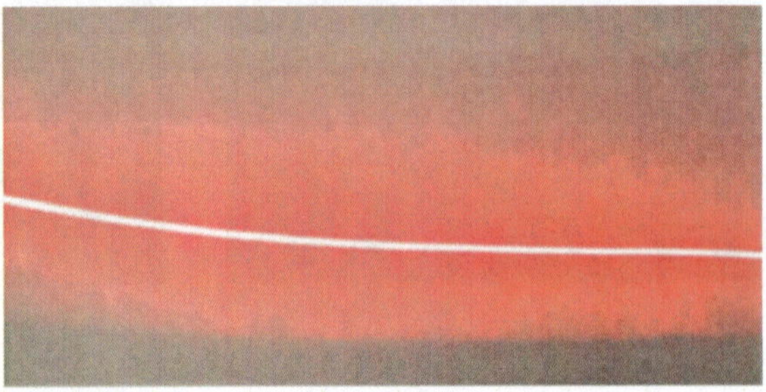

Fig. 11 Generation of striped structured light

by the laser passes through a cylindrical lens to become a linear laser beam, as shown in Fig. 11.

5 Conclusion

After the text edit has been completed, the paper is ready for the template. Duplicate the template file by using the Save As command, and use the naming convention prescribed by your conference for the name of your paper. In this newly created file, highlight all of the contents and import your prepared text file. You are now ready to style your paper; use the scroll down window on the left of the MS Word Formatting toolbar.

This paper proposes a fringe projection system for 3D measurement. The projection system consists of a MEMS micro-mirror, a controller, a drive circuit, a laser, and an optical module. The system has the advantages of simple principle and structure, easy operation and measurement. First of all, when choosing a linear MEMS surface, we know that the one-dimensional laser scanning module P1130 of micro-sensing has high accuracy, contains an electrostatically driven micro-mirror, is not easily affected by the external environment, and is suitable for a wide range of measurement locations. Red dot laser with TTL modulation signal, TTL modulation can facilitate us to find the most suitable output waveform; select the peripheral SPI of STM32F103C8T6 chip as the host of P1130, compile and debug the program with Keil uVision5, and connect to the personal computer through USB to TTL module Use FLYMCU to download the program to the chip, and finally output the angle data to generate striped structured light.

References

1. Guo HW (2001) PhD theses: 360° profilometry by multi-aperture overlap-scanning technique. Shanghai university
2. Saito K, Miyoshi T (1991) Development of non-contact 3-D digitizing and machining system for free-form surface. Annals CIRP 40(1):483–486
3. Kim WS (1999) Computer vision assisted virtual reality calibration. IEEE Trans Robot Autom 15(3):450–464
4. Chen F, Brown GM, Song M (2000) Overview of three-dimensional shape measurement using optical methods. Opt Eng 39(1):10–22
5. Yasser H (1994) Laser based system for reverse engineering. Comput Eng 26(2):387–394
6. Hepplewhite L, Stonham TJ, Glover RJ (1996) Automated visual inspection of magnetic disk media. In: International conference on electronics, circuits and system 732(2):732–735
7. He HT (2005) PhD theses: the research on 3D optical measurement techniques for complex shape. Shanghai university
8. Mu CJ (2008) Study on dynamic deformations of high-speed MEMS scanning micro-mirror. J Transducer Technol 21(4):640–643
9. Winter C, Fabre L, Conte F (2009) L: Micro-beamer based on MEMS micro-mirrors and laser light source. Proc Chem 1(1):1311–1314
10. Ma WY, Yao J, Ren H, Wang DJ (2009) Design and analysis of MEMS 2D electrostatic scanning mirrors. Micro-Nano Electron Technol 46(05):296–300

Promising Halide Perovskite: The Application in Field-Effect Transistors

Juanjuan Zhou, Yingquan Peng, and Wenli Lv

Abstract As the main component in electronic circuits, field-effect transistors (FETs) are indispensable in modern semiconductor industry. Thus, choosing appropriate functional layer materials has become an urgent matter to be solved. In recent years, halide perovskite is considered a favorable competitor with the continuously mature synthesis technologies. Especially, the effective light absorption and high charge carrier mobility make it possible to improve the performance of FETs. In this work, we summarized several representative halide perovskite-based FETs including thin-films, single crystals and mixed function layers. First, the perovskite films that can exhibit field-effect characteristics at low or normal temperature were introduced. Afterward, considering that there have no crystal boundaries in single crystals and mixed functional layers will be good for combining the advantages of different materials, typical corresponding devices were investigated in detail. Finally, the potential development of halide perovskite-based FETs in the future were outlined from three aspects.

Keywords Halide perovskite · Field-effect transistor · Field-effect characteristics

1 Introduction

As a promising candidate for optoelectronic materials, halide perovskite has attracted extensive attention due to the inherent tunable band gap, large absorption coefficient, extended electron–hole diffusion length and long carrier lifetime [1–4]. Especially for solar cells, the efficiency has rapidly increased to 22.1% by utilizing perovskite materials as light absorbing layer [5]. Meanwhile, they have also been widely used

J. Zhou · Y. Peng (✉) · W. Lv
College of Optical and Electronic Technology, China Jiliang University, Hangzhou 310018, P.R. China
e-mail: yqpeng@cjlu.edu.cn

Y. Peng and X. Dong (eds.), *Proceedings of 2019 International Conference on Optoelectronics and Measurement*, Lecture Notes in Electrical Engineering 726,
https://doi.org/10.1007/978-981-33-4110-4_26

in light-emitting diodes (LEDs), electrically pumped lasers and photodetectors [6–8] due to above-mentioned advantages. However, there has relatively less research effects in field-effect transistors (FETs) based on halide perovskites. According to existing literature, it may be mainly blamed for the gate-field screening effect caused by ion migration, which yield the mobility of halide perovskite thin-films unsatisfactory at room temperature. Set $CH_3NH_3PbI_3$ thin-film as an example, the mobility in light-emitting FET is as low as 10^{-4} cm^2/Vs at room temperature, which is far less than the first-principle calculation [9]. Yet this is not only a challenge but also an opportunity for the investigations of halide perovskite-based FETs if the issue of gate-field screening can be mitigated.

Here ignoring the gate-field screening issue, halide perovskite should be regarded as a promising candidate to develop advanced FETs. First, it can be synthesized easily from 0- to 3-dimension (D) as the synthesis technologies continue to mature, which is conducive to the application of FETs with different functional requirements. For the perovskites with a low dimensional geometry, the reduced defect density and improved carrier mobility will beneficial for harvesting charges that induced by photons [10]. Furthermore, the channel thickness of low dimensional perovskite structures is greatly reduced, which is crucial for obtaining smaller dark currents. Besides, it is worth mentioning that Deng et al. have fabricated the ultrahigh-mobility organic semiconductor single-crystal (OSSC) arrays by channel-restricted meniscus self-assembly (CRMS) method successfully [11]. That means the large-area arrays preparation has become possible, which will promote the development of multifunctional integrated circuits. At the same time, it can be integrated into other devices to get better performance. Second, the low-cost solution methods have been widely applied to halide perovskite thin-films preparation, which makes it possess strong processability. On the other hand, the processability of thin-films is beneficial to speed up the development of flexible and stretchable electronics, while the traditional silicon (Si)-based semiconductors even cannot achieve related applications. The final point is that the most common halide perovskite $CH_3NH_3PbI_3$ exhibits ambipolar characteristic without additional doping to balance the conductivity of n-and p-type.

Different from other reviews; i.e., Miao et al. presented the recent progress on highly sensitive perovskite photodetectors and Liu et al. investigated the synthesis, ion migration, and application based on metal halide perovskite in FETs [12, 13]. In this review, we specifically discuss the halide perovskite-based FETs. First, the current applications of pure thin-film halide perovskite-based FETs were introduced, in which several investigations put forward that the obvious field-effect characteristics of $CH_3NH_3PbI_3$ can be observed only at low temperature. While other investigations propose that this issue can be solved through optimization strategies such as the application of external light source. Afterwards, utilizing single crystals and mixed functional layers to improve the performance of halide perovskite-based FETs were enumerated respectively.

2 Thin-Film FETs

With the rapid development of flexible and stretchable devices, Si-based FETs have been unable to meet the needs of equipment production. However, halide perovskite thin-films can be easily obtained via a variety of solution methods, which makes them attracted considerable attention over the past decades. Labram et al. studied the FET employing $CH_3NH_3PbI_3$ as active material and conducted a series of temperature-dependent measurements [14]. As depicted in Fig. 1a, they found that the $CH_3NH_3PbI_3$-based FET exhibited no field-induced current modulation with low source-drain currents at room temperature. However, the field-effect began to appear when the temperature dropped to 220 K, and the drain current continued to increase with reduced temperature. It is concluded by them that there have no field-effect characteristics at room temperature is attributed to the gate-field screening effect, which

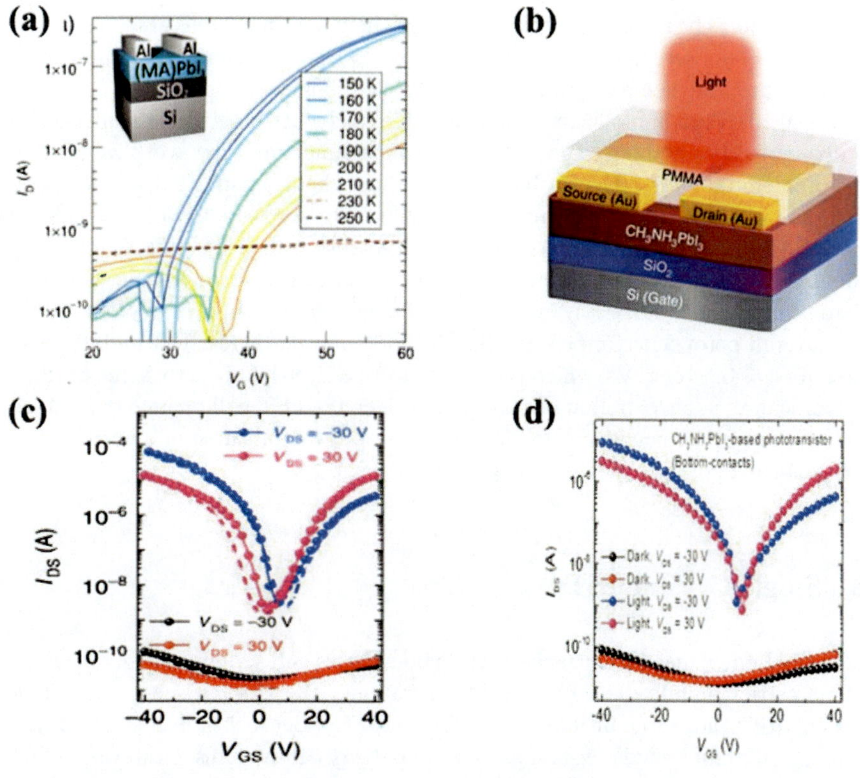

Fig. 1 **a** Transfer curves of $CH_3NH_3PbI_3$-based field-effect transistor (FET), which measured at a series of temperatures from 150 to 250 K. Reproduced with permission [14]. Copyright © 2015, American Chemical Society. **b** Schematic of the bottom-gate, top-contact FET based on CH3NH3PbI3 thin-film. Transfer characteristics of **c** bottom-gate, top-contact and **d** bottom-gate, bottom-contact devices in the dark (black and red symbols) and under light illumination (blue and magenta symbols). Reproduced with permission [16]. Copyright © 2015, Springer Nature

inhibits further charge carrier accumulation. Similar studies have also suggested that it is difficult to obtain the field-effect characteristics based on pure $CH_3NH_3PbI_3$ perovskite at room temperature [9, 15]. However, this situation is not inevitable.

For instance, Li et al. set an external light source as the fourth terminal electrode, which assisted the gate electrode to control channel current [16]. Ultimately, the $CH_3NH_3PbI_3$-based FET exhibited obvious room-temperature ambipolar characteristics. As shown in Fig. 1a, b bottom-gate, top-contact (BGTC) $CH_3NH_3PbI_3$-based FET was fabricated, in which the $CH_3NH_3PbI_3$ films was prepared by two-step vapour-assisted solution process. And poly (methyl methacrylate) (PMMA) was spin-coated as protective layer to prevent the diffusion of moisture and atmospheric oxygen. An ambipolar characteristic can be clearly observed in the transfer curves under light illumination as illustrated in Fig. 1c, in which the photo-induced hole and electron mobility are calculated as 0.18 and 0.17 cm^2/Vs respectively in saturation region. Hereafter, they also fabricated a bottom-gate, bottom-contact (BGBC) perovskite FET to determine whether the device structure influences the field-effect characteristic. The result was shown in Fig. 1d, in which an obvious ambipolar characteristic was observed to be like that of the BGTC device under light illumination. Furthermore, the electron and hole mobility were both measured as 0.17 cm^2/Vs, revealing that the ambipolar characteristic of $CH_3NH_3PbI_3$-based FETs is an inherent property and has nothing to do with the device architecture. This work well avoids the problem that $CH_3NH_3PbI_3$ has no field-effect characteristics at room temperature, and it has application significance in other optoelectronic devices. In addition to $CH_3NH_3PbI_3$ perovskite, 2D perovskites have also been used in FETs for a long time. Kagan et al. demonstrated a thin-film FET using 2D layered organic–inorganic perovskite $(C_6H_5C_2H_4NH_3)_2SnI_4$ as the semiconducting channel [17]. Note that the 2D layered perovskite formed a p-channel FET and the field-effect mobility was measured as 0.55 cm^2/Vs, which is different from $CH_3NH_3PbI_3$ perovskite. Furthermore, with the concentration of majority carriers, the FET will operate in accumulation mode under negative gate bias. In contrast, the application of a positive gate bias depleted holes in the channel, turning the device off.

3 Single Crystal FETs

It is well known that the existence of grain boundaries will seriously hinder the field-effect carrier mobility and FET device operations. Therefore, it seems that single crystals (SCs) are more suitable to be applied into FETs due to their less defectives and eliminated grain boundaries compared with polycrystalline films. However, in fact, there are almost no reports about SC perovskite-based FETs to date. It is speculated that the incomplete precursor conversion and hydration of SC perovskite faces lead to extensive surface contamination, thereby impairing the operation of FET devices.

Considering the above issues, Yu et al. demonstrated a spatially-confined inverse temperature crystallization strategy (as shown in Fig. 2a) to synthesize the controlled micrometer-thin single crystals (TSCs) of $CH_3NH_3PbX_3$ (X = Cl, Br, I) perovskite

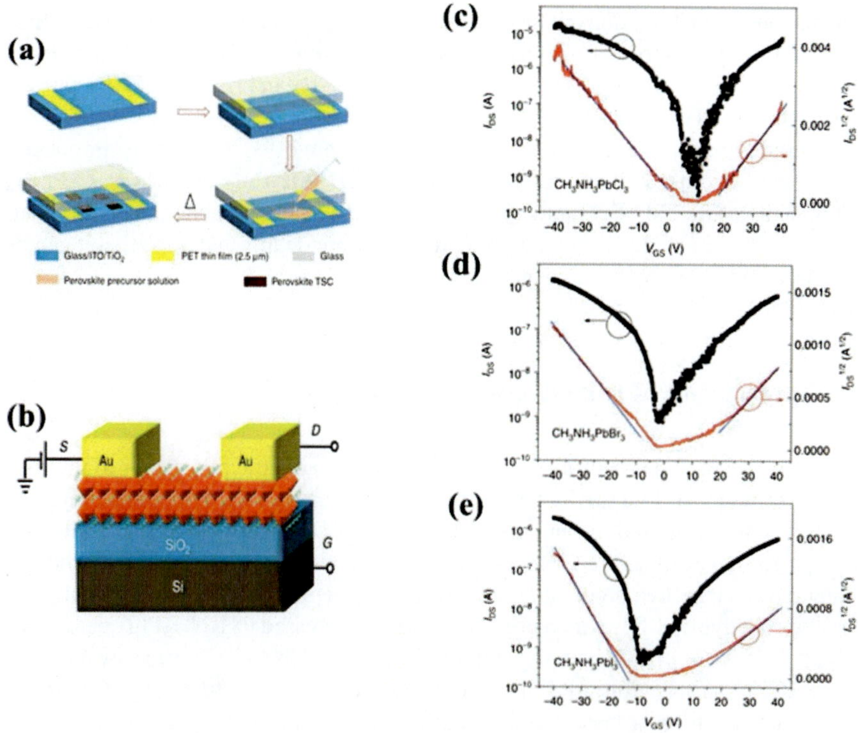

Fig. 2 **a** Schematic representation of spatially-confined inverse temperature crystallization strategy for producing controlled micrometer-thin single crystals (TSCs). **b** Schematic of the bottom-gate, top-contact (BGTC) TSC-FETs. Transfer characteristics, in which the black solid squares represent I_{DS} versus V_{GS}, the red solid lines represent I_{DS} versus V_{GS} and the blue solid line represent fit lines for **c** $CH_3NH_3PbCl_3$, **d** $CH_3NH_3PbBr_3$, and **e** $CH_3NH_3PbI_3$ TSC-FETs. Reproduced with permission [18]. Copyright © 2018, Springer Nature

[18]. Moreover, the grown TSCs possess sub-nanometer surface roughness and very low surface contamination, which benefit them to be integrated into ambipolar transistors. BGTC FETs with channel lengths (L) ranging from 10 to 150 μm were displayed in Fig. 2b, in which the channel widths (W) were controlled by the lateral size of TSCs. As shown in Fig. 2c–e, the fabricated devices exhibited obvious ambipolar transfer characteristics based on $CH_3NH_3PbX_3$ TSCs. It is worth mentioning that the maximum saturation hole (electron) mobilities were achieved to 2.6 (2.2), 3.1 (1.8), and 2.9 (1.1) cm^2/Vs for X = Cl, Br, and I, respectively. As a comparison, the BGBC FETs based on $CH_3NH_3PbX_3$ TSCs were fabricated by a similar method, achieving the maximum linear hole (electron) mobilities of 3.8 (0.32), 3.6 (0.26), and 4.7 (1.51) cm^2/Vs at room temperature for X = Cl, Br, and I, respectively. It is suggested that the successful demonstration of high-performance TSC-FETs at room temperature is attributed to the superior semiconductor-dielectric interfaces in the channel. In other words, this approach reduced the roughness,

defects, and contaminations in crystal face, eliminating the grain boundaries which reduced the barriers to lateral transport in FETs. Simultaneously, the vertical confinement made the crystals easy to grow laterally, so that it can bridge the FET channels ranging from 10 to 150 μm.

It should be annotated here that in addition to the above two forms, there are considerable several other halide perovskites such as $CH_3NH_3PbI_3$ nanowires, $CsPbBr_3$ nanoplatelets and $CsPbBr_3$ nanocrystals [10, 19, 20], showing potential applications in FETs are not listed. We are not going to give more details and hope that it can be investigated further by future generations.

4 Application of Mixed Functional Layer

Recently, the application of mixed function layers is an attractive means to improve the performance of FETs, which well fuses the properties of different materials. For instance, amorphous indium gallium zinc oxide (IGZO) is worthy of consideration for high performance thin-film TFTs, due to its high field-effect carrier mobility and low temperature large area processing capabilities. As depicted in Fig. 3a, Du et al. have developed a hybrid FET, which utilized solution-processed $CH_3NH_3PbI_3$ perovskite to cap IGZO [21]. Interestingly, the spin-coated $CH_3NH_3PbI_3$ film covering the entire IGZO surface appeared in a needle-like network structure (Fig. 3b), which was optimized to guarantee low dark current. Furthermore, the calculated responsivity (*R*) were plotted in Fig. 3c, which indicated the high sensitivity to light with $CH_3NH_3PbI_3$/IGZO FET in both ultraviolet and visible regions. In detail, both $CH_3NH_3PbI_3$/IGZO and IGZO FETs reached the maximum *R* values at ultraviolet range ($\lambda = 385$ nm). While with a broad absorbance of $CH_3NH_3PbI_3$ perovskite, the *R* values enhanced much for $CH_3NH_3PbI_3$/IGZO FET compared with that of IGZO FET in visible range ($\lambda = 550$–750 nm).

In the same year, Yu et al. proposed that the FETs which use $CH_3NH_3PbI_3$ perovskite-PbSe colloidal quantum dots heterostructures as functional layer exhibited wide spectral characteristic ranging from 300 to 1500 nm [22]. Furthermore, the responsivities at different wavelengths maintained in a high level due to the combination of perovskite and quantum dots, especially at 460 nm up to 1.2 A/W. Therefore, these results indicated that halide perovskites can combine their advantages with other materials well, so that more specific devices needed by people may be fabricated.

5 Conclusion and Outlook

In this paper, several different types of typical halide perovskite-based field-effect transistors (FETs) have been summarized and discussed exhaustively. Irrespective of

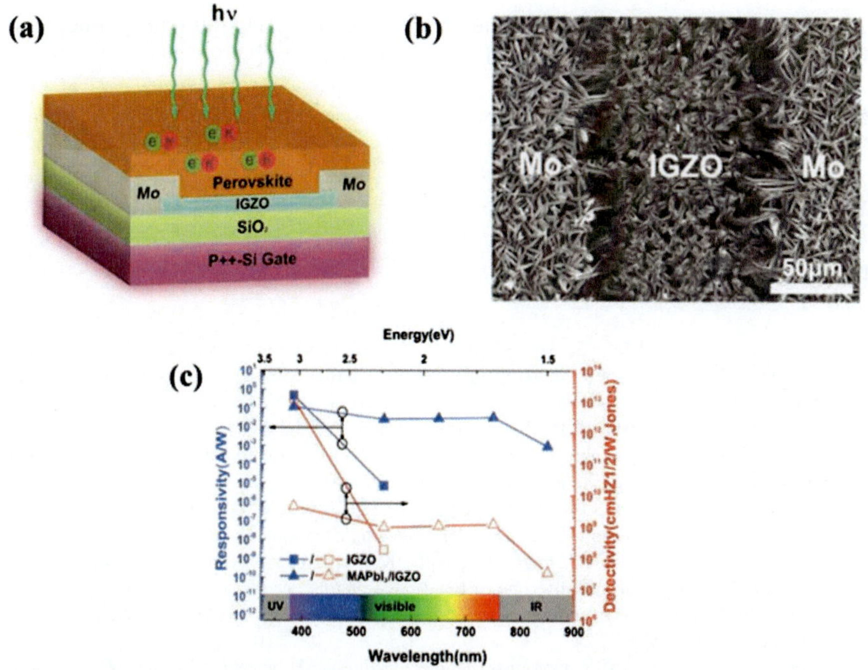

Fig. 3 **a** Schematic of $CH_3NH_3PbI_3$/IGZO hybrid FET. **b** The optical image taken from scanning electron microscopy (SEM) of CH3NH3PbI3 layer, which located in the channel region. **c** Responsivity of IGZO and CH3NH3PbI3/IGZO FETs at VG = − 10 V. Reproduced with permission [21]. Copyright © 2017 WILEY-VCH Verlag GmbH & Co. KGaA, Weinheim

gate-field screening issue, the halide perovskites are considerable materials to fabricate high-performance FETs due to the inherent outstanding optoelectronic properties: large absorption coefficient, extended electron–hole diffusion length, and tunable band gap. Thin-film is the most common morphological structure, but it has been suggested by several groups that the field-effect characteristics based on $CH_3NH_3PbI_3$ can only be observed at low temperatures. However, it can be solved by new strategies such as the addition of an external light source to set the fourth terminal electrode. Furthermore, two-dimension perovskite can also obtain field-effect characteristics at room temperature. In terms of single crystals, a spatially-confined inverse temperature crystallization strategy can be used to promote the performance of FETs. The most interesting is that the combination of halide perovskites and other materials such as IGZO and PbSe are of great help for broadband detection and so on.

It can be predicted that the halide perovskite thin-films will fill the vacancy of conventional Si-based FETs in flexible and stretchable electronics because of its processability. Moreover, with the continuous development of single crystal synthesis technology, once the problem of incomplete precursor conversion and hydration in SC perovskite faces is solved, it will shine in integrated circuits. In addition to the application of mixed functional layers, we believe that through in-depth research and new strategy development of halide perovskite, more and more multifunctional field-effect transistors will be applied to various fields.

Acknowledgements This work was supported by Natural Science Foundation of Zhejiang Province Grant No. LY18F050009, and National Key R&D Program of China Grant No. 2016YFF0203605.

References

1. Burschka J, Pellet N, Moon SJ, Humphry-Baker R, Gao P, Nazeeruddin MK, Grätzel M (2013) Sequential deposition as a route to high-performance perovskite-sensitized solar cells. Nature 499(7458):316–319
2. Sun SY, Salim T, Mathews N, Duchamp M, Boothroyd C, Xing GC, Sum TC, Lam YM (2014) The origin of high efficiency in low-temperature solution-processable bilayer organometal halide hybrid solar cells. Energy Environ Sci 7(1):399–407
3. Dong QF, Fang YJ, Shao YC, Mulligan P, Qiu J, Cao L, Huang JS (2015) Electron-hole diffusion lengths > 175 μm in solution-grown $CH_3NH_3PbI_3$ single crystals. Science 347(6225):967–970
4. Chen T, Chen WL, Foley BJ, Lee J, Ruff JPC, Ko JY, Brown P, Harriger CM, Zhang LM, Lee D (2017) Origin of long lifetime of band-edge charge carriers in organic-inorganic lead iodide perovskites. In: Proceedings of the national academy of sciences of the United States of America 114(29):7519–7524
5. Saliba M, Matsui T, Seo JY, Domanski K, Correa-Baena JP, Nazeeruddin MK, Zakeeruddin SM, Tress W, Abate A, Hagfeldt A, Grätzel M (2016) Cesium-containing triple cation perovskite solar cells: improved stability, reproducibility and high efficiency. Energy Environ Sci 9(6):1989–1997
6. Song JZ, Fang T, Li JH, Xu LM, Zhang FJ, Han B, Shan QS, Zeng HB (2018) Organic-inorganic hybrid passivation enables perovskite QLEDs with an EQE of 16.48% 30(50):1805409
7. Zhu HM, Fu YP, Meng F, Wu XX, Gong ZZ, Ding Q, Gustafsson MV, Trinh MT, Jin S, Zhu XY (2015) Lead halide perovskite nanowire lasers with low lasing thresholds and high quality factors. Nat Mater 14(6):636-U115
8. Fang YJ, Dong QF, Shao YC, Yuan YB, Huang JS (2015) Highly narrowband perovskite single-crystal photodetectors enabled by surface-charge recombination. Nat Photon 9(10):679–687
9. Chin XY, Cortecchia D, Yin J, Bruno A, Soci C (2015) Lead iodide perovskite light-emitting field-effect transistor. Nat Commun 6:7383
10. Li TT, Liu MH, Li QY, Chen R, Liu X (2018) Hybrid photodetector based on CsPbBr 3 perovskite nanocrystals and P_{C7}1BM fullerene derivative. Chem Phys Lett 699:208–211
11. Deng W, Zhang XJ, Dong HL, Jie JS, Xu XZ, Liu J, He L, Xu L, Hu WP, Zhang XH (2019) Channel-restricted meniscus self-assembly for uniformly aligned growth of single-crystal arrays of organic semiconductors. Mater Today 24:17–25
12. Miao JL, Zhang FJ (2019) Recent progress on highly sensitive perovskite photodetectors. J Mater Chem C 7(7):1741–1791
13. Liu XH, Yu DJ, Song XF, Zeng HB (2018) Metal halide perovskites: synthesis, ion migration, and application in field-effect transistors. Small 14(36):1801460
14. Labram JG, Fabini DH, Perry EE, Lehner AJ, Wang HB, Glaudell AM, Wu G, Evans H, Buck D, Cotta R, Echegoyen L, Wudl F, Seshadri R, Chabinyc ML (2015) Temperature-dependent polarization in field-effect transport and photovoltaic measurements of methylammonium lead iodide. J Phys Chem Lett 6(18), 3565–3571
15. Li DH, Cheng HC, Wang YL, Zhao ZP, Wang GM, Wu H, He QY, Huang Y, Duan XF (2017) The effect of thermal annealing on charge transport in organolead halide perovskite microplate field-effect transistors. Adv Mater 29(4):1601959

16. Li F, Ma C, Wang H, Hu WJ, Yu WL, Sheikh AD, Wu T (2015) Ambipolar solution-processed hybrid perovskite phototransistors. Nat Commun 6:8238
17. Kagan CR, Mitzi DB, Dimitrakopoulos CD (1999) Organic-inorganic hybrid materials as semiconducting channels in thin-film field-effect transistors. Science 286(5441):945–947
18. Yu WL, Li F, Yu LY, Niazi MR, Zou YT, Corzo D, Basu A, Ma C, Dey S, Tietze ML, Buttner U, Wang XB, Wang ZH, Hedhili M, Guo NC, Amassian A (2018) Single crystal hybrid perovskite field-effect transistors. Nat Commun 9:5354
19. Spina M, Náfrádi B, Tóháti HM, Kamarás K, Bonvin E, Gaal R, Forró L, Horváth E (2016) Ultrasensitive 1D field-effect phototransistor: $CH_3NH_3PbI_3$ nanowire sensitized individual carbon nanotube. Nanoscale 8:4888–4893
20. Huo CX, Liu XH, Song XF, Wang ZM, Zeng HB (2017) Field-effect transistors based on van-der-Waals-grown and dry-transferred all-inorganic perovskite ultrathin platelets. J Phys Chem Lett 8(19):4785–4792
21. Du SN, Li GT, Cao XH, Wang Y, Lu HL, Zhang SD, Liu C, Zhou H (2017) Oxide semiconductor phototransistor with organolead trihalide perovskite light absorber. Adv Electron Mater 3(4):1600325
22. Yu Y, Zhang YT, Zhang Z, Zhang HT, Song XX, Cao MX, Che YL, Dai HT, Yang JB, Wang JL, Zhang H, Yao JQ (2017) Broadband phototransistor based on $CH_3NH_3PbI_3$ perovskite and PbSe quantum dot heterojunction. J Phys Chem Lett 8(2):445–451

Application of Advanced Optic Measurement Technology DISAR in Engineering Modeling and Simulation

Lan Wei and Miao Zheng

Abstract The testing technology displacement interferometer system for any reflector (DISAR) is a kind of advanced optic measurement technique, which has been widely used in many fields, such as detonation physics, impact dynamics, and so on. In this paper we adopted DISAR testing to measure the high-speed movement process of detonation driven metal flyer. Compared with the traditional electric probe testing methods, DISAR testing can effectively ensure the testing accuracy in along with the increasing of measurement range. Then we carried out modeling and simulation on the movement of metal flyer driven by IHE utilizing a two-dimensional Lagrange finite element code. According to the velocity-time curve obtained by DISAR testing, we modified the mesh ratio and the material model of metal flyer. The modified mathematical method can well simulate the velocity history of flyer. Results show that the advanced optic measurement technique DISAR can play an important role in engineering modeling and simulation.

Keywords Advanced Optic Measurement Technology · DISAR · Velocity History · Modeling and Simulation

1 Introduction

Velocity interferometer system for any reflector (VISAR) [1, 2] and displacement interferometer system for any reflector (DISAR) [3, 4] are both advanced optic measurement techniques for the continuous testing of surface velocity, based on the basic principle of optical Doppler effect, using optical fibers to transmit light signal, and using the optical interference mixing technology to obtain the continuous velocity information of an object. The principles of VISAR and DISAR are slightly different. Both of the two coherent beams of VISAR have frequency shift signal,

L. Wei · M. Zheng (✉)
Institute of Applied Physics and Computational Mathematics, Beijing 100094, P. R. China
e-mail: zheng_miao@iapcm.ac.cn

Y. Peng and X. Dong (eds.), *Proceedings of 2019 International Conference on Optoelectronics and Measurement*, Lecture Notes in Electrical Engineering 726,
https://doi.org/10.1007/978-981-33-4110-4_27

but there is time delay. One of the two coherent beams of DISAR has no frequency shift signal and the other has frequency shift signal. Compared with VISAR, which is sensitive to the number of total interference fringes [5], DISAR is not sensitive to the number of total interference fringes, and the loss of the number of total interference fringes has little effect on the measurement. Currently DISAR testing has been widely used in many fields, such as detonation physics, impact dynamics, geological science, material science, medical diagnosis, and so on [6–11].

In the present study we adopted DISAR testing to measure the high-speed movement process of metal Fe flyer driven by TATB-based insensitive high explosives (IHE) detonation and compared with the traditional electric probe testing methods. We also carried out modeling and simulation on the movement of metal flyer driven by IHE utilizing a two-dimensional Lagrange finite element code. According to the velocity-time curve obtained by DISAR testing, we modified the mesh ratio and the material model of metal flyer.

2 Experiment

2.1 Principle of Testing

The basic principle of laser interferometer testing is as follow [12]. The laser emits a beam to the surface of object and the frequency of light received by the detect equipment can be expressed by Equation (1).

$$f = f_0 + \frac{v}{c} \cdot f_0 \cdot (\cos\theta_1 + \cos\theta_2) \quad (1)$$

Where f and f_0 are frequencies of reflecting light and entering light, θ_1 and θ_2 are the angles formed by the object moving direction and the entering light and reflecting light, v is the velocity of object, and c is the light speed in the air.

In DISAR system, both the entering light and the reflecting light are vertical to the surface of object, so the frequency change of reflecting light according to entering light can be simplified as Equation (2).

$$\Delta f = f - f_0 = \frac{2v}{c} \cdot f_0 = \frac{2v}{\lambda_0} \quad (2)$$

Where λ_0 is the wave length of entering light.

The diagram of DISAR System is shown in Fig. 1 [13]. The light wave emitted by the laser is divided into two beams. The first beam directly enters the fiber coupler as the local light. The second beam is projected on the target surface, reflected into the amplifier, and then into the fiber coupler as the detection light. At last, two interference signals are output by the fiber coupler and recorded by the photo detector and the oscilloscope respectively.

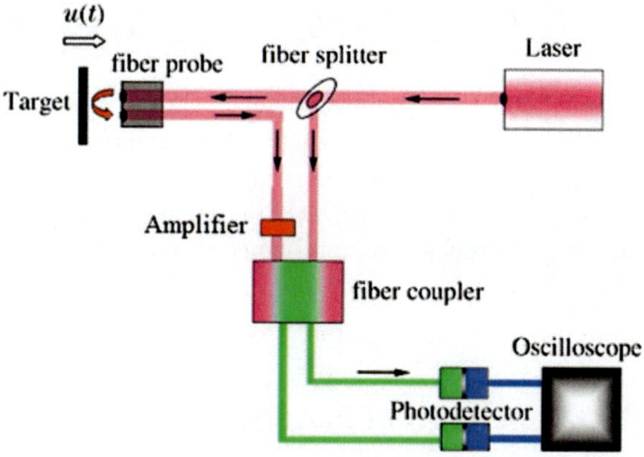

Fig. 1 The diagram of the DISAR system

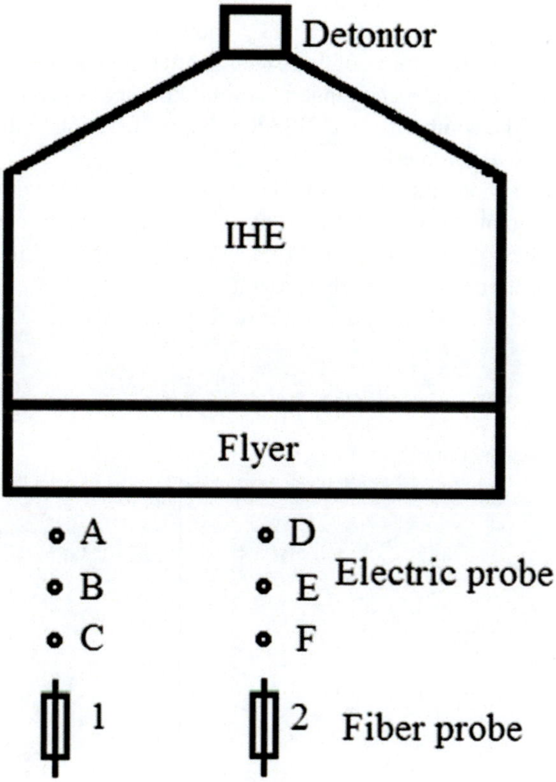

Fig. 2 The diagram of the experimental set

2.2 Experimental Set

The diagram of experimental set is shown in Fig. 2. The Fe flyer was driven by TATB-based IHE of a designated shape and electric probes and DISAR fiber probes were placed at different points of the space.

Arriving times of flyer at different points A, B, C, D, E and F were tested by the electric probes and continuous velocity changes of flyer at certain angles were obtained by DISAR fiber probes 1 and 2.

2.3 Experimental Results

Figure 3 shows two velocity curves obtained by DISAR optic probes 1 and 2. Physical quantities including time, distance and velocity given in this paper are all dimensionless, for we focus on the application effect of the testing technology, not the value of experimental data. Figure 3 shows that DISAR testing can obtain a continuous velocity-time curve of flyer, with an effective testing time for dozens of microseconds, and a testing speed range of several kilometers per second.

The comparison of DISAR velocity curve integration and electric probes data is shown in Table 1.

By integrating velocity curve 1 in Fig. 3, we can obtain arriving times of flyer at points A, B, C. Similar, by integrating velocity curve 2, we can obtain arriving times of flyer at points D, E and F. The arrival time of the flyer at characteristic location obtained by the integration of velocity-time curve is accord with the result by electric probe testing and the difference is less than 0.3%, which shows that DISAR

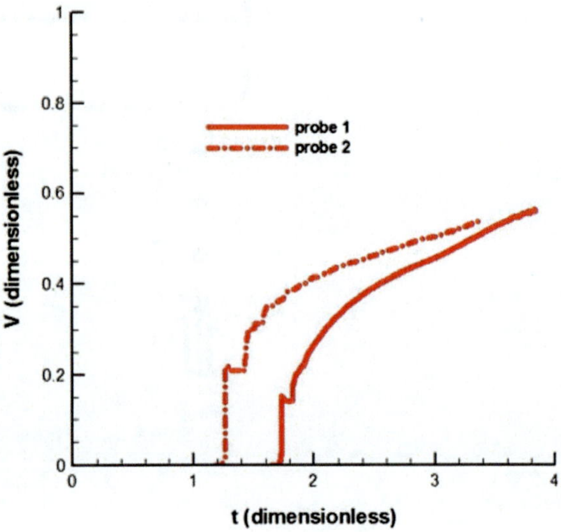

Fig. 3 The velocity of two DISAR fiber probes as a function of time

Table 1 Comparison of DISAR and electric probes

Position	Electric probe	Integration of DISAR curve	Difference(%)
A	3.453	3.447	0.19
B	3.740	3.747	0.18
C	3.980	3.987	0.17
D	2.473	2.468	0.27
E	2.867	2.860	0.23
F	3.233	3.227	0.21

testing can effectively ensure the testing accuracy in along with the increasing of measurement range.

3 Modeling and Simulation

3.1 Models and Parameters

We carried out modeling and simulation on the movement of metal flyer driven by IHE utilizing a two-dimensional Lagrange finite element code. We used detonation Shock Dynamics (DSD) method to describe the detonation propagation of TATB-based IHE and the Jones-Wilkins- Lee (JWL) EOS as Equation (3) to describe the detonation products. The ideal elastic-plastic constitutive model was used to Fe flyer.

$$p = A(1 - \frac{\omega}{R_1 v})e^{-R_1 v} + B(1 - \frac{\omega}{R_2 v})e^{-R_2 v} + \omega E/v \tag{3}$$

Where p is pressure, v is relative volume, E is relative inner energy, ω is the Gruneisen coefficient, and A, B, R_1, and R_2 are constants.

3.2 Results and Discussion

Figure 4 shows the comparison of calculated velocity history curve with the DISAR testing data. The tendency of the calculated velocity history is similar to the experiment, but the first jump and the following segment of the calculated velocity is obviously different from the experimental result.

According to the velocity-time curve obtained by DISAR testing, we modified the modeling and simulation method. On the one hand, the mesh ratio of the metal flyer was changed from three to one. On the other hand, the constitutive model of metal material was modified from ideal elastic-plastic model to SG model. The comparison

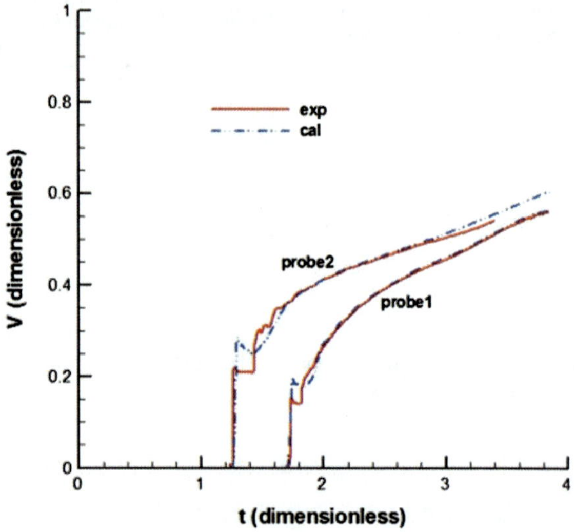

Fig. 4 The comparison of calculated velocity curve with the experimental data

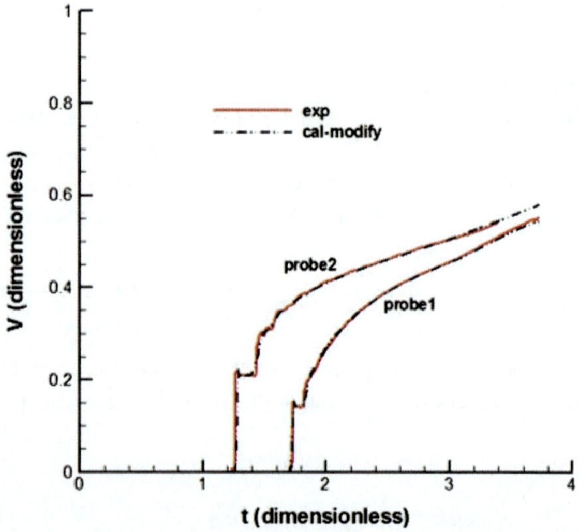

Fig. 5 The comparison of modified calculated velocity curve with the experimental data

of modified calculated velocity curve with the experimental data is shown in Fig. 5. Both the first jump and the following parts of the curve agree well with the experiment.

4 Conclusions

We present our work on the application of advanced optic measurement technology DISAR in engineering modeling and simulation. The surface velocity curve of Fe flyer driven by TATB-based IHE detonation was obtained by DISAR. DISAR testing can effectively ensure the testing accuracy in along with the increasing of measurement range. According to the velocity-time curve obtained by DISAR testing, we modify the modeling and simulation method, and the modified mathematical method can well simulate the velocity history of flyer. Results show that the advanced optic measurement technique DISAR can play an important role in engineering modeling and simulation.

Acknowledgement This work was supported by the Foundation of 2018-JCJQ-ZQ-028.

References

1. Wang SY, Cao CZ, Zhao XL (1986) Principles of a laser velocity interferometer (VISAR) and some experimental results. ACTA Armamentaril 4:41–45
2. Zhao JH, Sun CW, Ma RC, Duan ZP (2000) Development and application of VISAR probe with higher signal collecting efficiency and adjusted depth of field. J Exp Mech 15(1):1–8
3. Mercier P, Benier J, Azzolina A et al (2006) Photonic Doppler velocimetry in shock physics experiments. J Phys IV France 134:805–812
4. Tasker DG, Whitley VH, Lee RJ (2009) Electromagnetic field effects in explosives. In: 16th APStopical conference on shock compression of condensed matter 54(8)
5. Li ZR, Yao JQ (1999) Signal loss and determination of lost fringe number in VISAR application. Explosion Shock Waves 19(2):182–187
6. Li XM, Wang XS, Wang PL, Lu M, Jia LF (2009) Spall of cylindrical copper by converging sliding detonation. Explosion Shock Waves 29(2):162–166
7. Matthew EB, Larry GH, Lawrence MH et al (2010) Applications and principles of photon doppler velocimetry for explosive testing. In: 14th international symposium on detonation. Coeurd' Alene, USA
8. Qin WZ, Long XP, He B, Jiang XH (2012) Experiments and numerical simulation of flyers driven by BNCP. Initiators Pyrotech 1:15–17
9. Yu YY, Xi F, Dai CD, Cai LC, Tan H, Li XM (2014) Measurement of strength in a Zr-based bulk metallic glass under dynamic high-pressure loading. Explosion Shock Waves 34(1):1–5
10. Zheng ZJ, Duan ZP, Zhang LS, Ou ZC, Huang FL (2015) Experimental research on the dynamic response of the Al2O3 ceramic at impact loading. Trans Beijing Institute Technol 35(8):792–795
11. Shi GK, Yang J, Liu WX, Zhang M, Wang Z, Xu HB, Tang SY, Hu HQ, Hui HL (2016) Photonic doppler velocimetry of outer layer of sand wall under internal explosion loading. China Measurem Test 42(10):68-71
12. Wang X, Tan KY, Wen SG, Liu QJ, Ye H (2014) Influence of flyer materials on flyer velocity driven by electric explosion. Chinese J Energ Mater 22(2):259–262
13. Weng JD, Li YL, Chen H, Ye XP, Ye SH, Tan H, Liu CL (2018) Application of an all-fiber displacement interferometer on SHPB experiment measurements. Chinese J High Pressure Phys 32(1):013201-1-6

The Design of Integrated Four-Channel Mach–Zehnder Multi/Demultiplexer Based on LNOI Platform

Liang Xia, Shiqi Tao, Cheng Zeng, and Jinsong Xia

Abstract As a kind of platform for optoelectronic integrated circuits, lithium niobate on insulator (LNOI) possesses excellent material properties of lithium niobate. With a better optical confinement contrast with other platforms, LNOI can improve the device integration greatly. We have designed a four-channel of filter based on the Mach–Zehnder structure on the LNOI. The simulation result shows that the insertion loss and cross talk of the filter are less than 1 dB and -10 dB, respectively. The filter can meet the requirement of CWDM4 specification for the next generation optical module and optical interconnects.

Keywords LNOI platform · CWDM4 · Multi/demultiplexer

1 Introduction

With the explosive growth of data traffic in the global scope, the demand for high-speed modules increases rapidly. Especially, with the application of lithium niobate on insulator (LNOI) platform, higher modulate rate such as 100 or 400 GHz is promising. LNOI is a kind of platform for optoelectronic integrated circuits, which not only possesses excellent material properties of lithium niobate, but also has a better optical confinement contrast with other platforms like SOI. These characteristics are expected to be applied to the preparation of high speed modules [1]. Besides, WDM devices can realize multichannel transmission and thereby increase the performance bandwidth, simultaneously reduce the transmission cost [2].

We have designed a four-channel of filter based on the Mach–Zehnder structure on the platform of LNOI, which conforms to the CWDM4 specification [3]. The device consisted of three filters based on the same Mach–Zehnder type. The only

L. Xia · S. Tao · C. Zeng · J. Xia (✉)
Wuhan National Laboratory for Optoelectronics, Huazhong University of Science and Technology, Wuhan 430074, China
e-mail: xlsd2016@hust.edu.cn

Y. Peng and X. Dong (eds.), *Proceedings of 2019 International Conference on Optoelectronics and Measurement*, Lecture Notes in Electrical Engineering 726,
https://doi.org/10.1007/978-981-33-4110-4_28

difference between them is the length difference L, which determines the central wavelength of the filter. Every filter is composed of three parts.The first part is a 3 dB coupler which is used for average power assignment. The second part is Mach–Zehnder interferometer which is usually used to provide phase difference. The third part is another type of coupler with two input ports and output ports.

2 Design Process

Multimode Interference Coupler is a type of coupler composed by input waveguide, multimode waveguide and output waveguide. Compared with directional coupler, multimode interference coupler has a better wavelength sensitivity, means average power assignment can be achieved within the demanded wavelength range. Therefore, we decide to use the multimode interference coupler as the first part. With the basic theory of self-imaging effect, the design parameters are determined and the simulation result is showed below. The result indicates that multimode interference coupler with the design parameters can meet the requirement for the filter. Besides, we use the directional coupler for the third part due to its smaller loss contrast with the same type of multimode interference coupler. (Fig. 1).

Mach–zehnder Interferometer is generally composed of two waveguide arms named delay waveguide with different length. Because of the length difference of the delay waveguide, the incident optical field will accumulate different phases in each arm when it passes through the interferometer.The length difference can be calculated by the following Formula [4]:

$$FSR = \frac{\lambda^2}{n_{gr}\Delta L} \tag{1}$$

Fig. 1 The schematic layout of the proposed CWDM (de)multiplexer

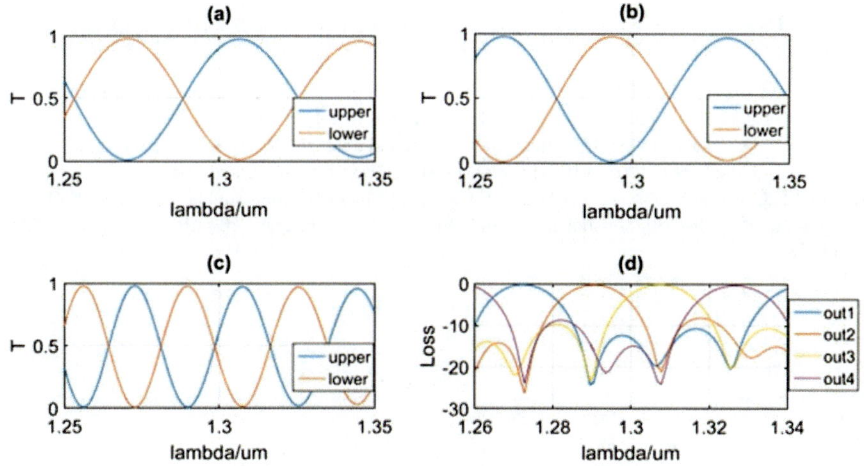

Fig. 2 **a** The simulation result of the first part to separate odd band from even band. **b** ~ **c** The simulation results of the second part and the third part. **d** The overall performance of the device

3 Simulation Results

The device consists of three filters, The first filter is used to separate the odd band from the even band. That means the upper waveguide will output Lam1 and Lam3, the lower waveguide will output Lam2 and Lam4, as shown in Fig. 2a. The second filter is used to separate Lam1 and Lam3, so the upper waveguide corresponds to Lam1 while the lower waveguide corresponds to Lam3(Fig. 2b). The third filter is used to separate Lam2 from Lam4, with the upper waveguide corresponding to Lam2, the lower waveguide corresponding to Lam4 (Fig. 2c).

We can find that the loss of the device is about 1 dB, simultaneously the crosstalk of the device is about −10 dB. This result means that the performance of the device can be further improved.The reason caused a large crosstalk is that the directional coupler cannot achieve wavelength uniformity which means the device should have the same splitting ratio to different wave length.

4 Outlook

The method involved in the design process is also suitable for LAN WDM which is another mature WDM technology. Besides, this device is designed and fabricated on LNOI platform, is a promising solution for high-speed modulator, so it is significance for monolithic integrated modulator which will be applied for modules in the future.

References

1. Boes A, Corcoran B, Chang L et al (2018) Status and potential of lithium niobate on insulator (LNOI) for photonic integrated circuits. Laser Photon Rev 1700256
2. Madsen CK, Zhao JH (1999) Optical filter design and analysis. Wiley Inc., New Jersey
3. Lewis D, Cole C et al (2017) 100G CWDM4 MSA Technical Specifications
4. Horst F, Green WM, Assefa S et al (2013) Cascaded Mach-Zehnder wavelength filters in silicon photonics for low loss and flat pass-band WDM (de-)multiplexing. Opt Expr 21(10):11652–11658

Design of Resonant Magnetic Field Sensor Based on Magnetostrictive Optical Fiber Micro-cantilever

Cheng Xu, Yueming Liu, and Yuchan Liu

Abstract In this paper, a magnetostrictive-based optical fiber micro-cantilever resonant magnetic field sensor is proposed. The magnetic field sensor is based on the optical fiber end face design of the optical fiber micro-cantilever beam. The surface of the optical fiber micro-cantilever beam is plated with a magnetostrictive film, and the two form a double layer micro-cantilever beam structure. The magnetostrictive film generates a magnetostrictive effect under a magnetic field, which causes the double-layer cantilever structure to deflect and change its resonant frequency. The magnetic field can be determined by detecting the change in resonant frequency. Then use ANSYS simulation software to simulate the resonance frequency of the double-layer micro-cantilever structure under the magnetic field, and obtain the relationship between the magnetic field and the resonance frequency, in order to optimize the size of the double-layer cantilever structure, and then obtain the best sensitivity of the magnetic field sensor. The simulation results show that: when the double-layer micro-cantilever structure is 90 μm long, the thickness of the fiber-optic micro-cantilever is 2 μm, and the thickness of the magnetostrictive film is 3/5 of the thickness of the fiber-micro-cantilever, the magnetic field sensor can reach the maximum sensitivity of 40,760 Hz/Gs.

Keywords Optical Fiber Micro-cantilever · Mgnetostrictive Film · ANSYS · Resonance

1 Introduction

Magnetic field sensors are devices that convert magnetic fields and their changes into other signals and output them, which is closely related to human life and has a wide range of applications in production, scientific research, and military. A traditional

C. Xu · Y. Liu (✉) · Y. Liu
China Jiliang University, HangZhou, China
e-mail: 07a0402054@cjlu.edu.cn

Y. Peng and X. Dong (eds.), *Proceedings of 2019 International Conference on Optoelectronics and Measurement*, Lecture Notes in Electrical Engineering 726,
https://doi.org/10.1007/978-981-33-4110-4_29

magnetic field sensor is a device that converts a magnetic field and its variation into an electrical signal output. For example, magnetic flux gages, Hall sensors, etc. Magnetic flux gages use the non-linear relationship between magnetic induction and magnetic field strength to measure the magnetic field under the saturation excitation of a high-permeability magnetic core in the measured magnetic field. The magnetic field is measured using a linear relationship between the Hall potential and the magnetic field. And these traditional magnetic field sensors use electric current when transmitting electrical signals. The magnetic effect of the generated current will interfere with the magnetic field to be measured, causing the sensor to generate errors when measuring the magnetic field. The optical fiber magnetic field sensor overcomes this shortcoming. Using the characteristics of all-optical fiber transmission, the magnetic field signal is converted into an optical signal, which eliminates the interference of the magnetic effect of the current on the magnetic field detection. The sensor probe is made on the sensor, so that the sensor is miniaturized, and the magnetic field in small and narrow places can be measured.

In this paper, a magnetostrictive-based optical fiber micro-cantilever resonant magnetic field sensor is proposed. The magnetic field sensor is based on the optical fiber end face design of the optical fiber micro-cantilever beam. The surface of the optical fiber micro-cantilever beam is plated with a magnetostrictive film, and the two form a double layer micro-cantilever beam structure. The magnetic field sensor uses a double-layer micro-cantilever structure as a magnetic field sensing element.The magnetostrictive film generates a magnetostrictive effect under a magnetic field, which causes the double-layer cantilever structure to deflect and change its resonant frequency. The magnetic field can be determined by detecting the change in resonant frequency, of the double-layer micro-cantilever structure in the magnetic field sensor, the resonance of the double-layer micro-cantilever structure under the magnetic field was simulated, and the structure was optimized to obtain higher sensitivity.

2 Resonance Theoretical Model of Double-Layer Micro-cantilever

Cantilever structure is a widely used structure in sensors, which has been used in biological, chemical, acceleration, humidity, force, heat and other sensors. When used as a sensor, it has two working modes: static and resonant. For the resonant mode, the resonant frequency of the cantilever beam is usually changed based on factors such as surface stress and mass. The working principle of the magnetic field sensor proposed in this article is: the double-layered micro-cantilever structure plated with a magnetostrictive film is affected by the magnetic field, and the magnetostrictive effect of the upper layer of the magnetostrictive film causes the magnetostrictive film and the optical fiber micro-cantilever. There is a mismatch between the stresses [1], which causes the double-layered micro-cantilever structure to deflect, which causes the resonance frequency of the double-layered micro-cantilever structure to change,

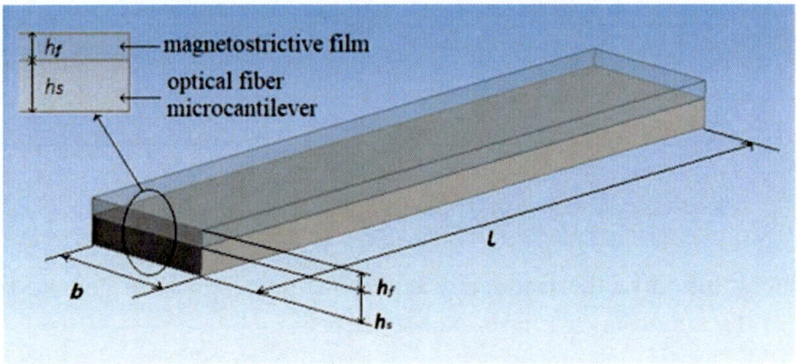

Fig. 1 The structure of double-layer micro-cantilever beam

and the magnetic field can be detected by detecting the change of the resonance frequency.

Based on the important role of the double-layer micro-cantilever structure, we analyze the resonance of the double-layer micro-cantilever structure under the influence of magnetic field. The structure of a double-layer micro-cantilever beam is shown in Fig. 1. The double-layer micro-cantilever structure has a length of l and a width of b; the thickness of the optical fiber micro-cantilever is h_s, the density is ρ_s, the Young's modulus is E_s, and the Poisson ratio is v_s; The thickness of the magnetostrictive film is h_f, the density is ρ_f, the Young's modulus is E_f, and the Poisson's ratio is v_f. Ignoring the effects of gravity and other factors, when the double-layer micro-cantilever structure is not affected by the magnetic field, the magnetostrictive film does not produce a magnetostrictive effect, and the double-layer cantilever beam remains straight.

First, we discuss the resonant frequency of a single-layer micro-cantilever. The resonance frequency of a uniform, flat and thin cantilever beam can be solved by the one-dimensional Euler–Bernoulli differential equation. For a micro-cantilever with a length of l, a width of b, and a thickness of h, the differential equation of motion is:

$$\tilde{E}I\frac{\partial^4 z(x,t)}{\partial x^4} = -\rho A\frac{\partial^2 z(x,t)}{\partial t^2} \tag{1}$$

where $\tilde{E} = E/(1 - v^2)$,$\tilde{E}$ is Biaxial tensile Young's modulus [2-3]; I is moment of inertia, $I = bh^3/12$;A is the cross-sectional area of a cantilever beam, $A = bh$;t is time; ρ is density. The boundary conditions of the cantilever beam are:

$$z(x=0) = z'(x=0) = z''(x=l) = z'''(x=l) = 0 \tag{2}$$

Solving Eq. (1) with Separated Variable Method, Let z (x, t) = Z (x) T (t), and bring it into Eq. (1) to get:

$$\beta^4 = \frac{\rho A \omega^2}{\tilde{E} I} \tag{3}$$

Bringing the boundary conditions into Eq. (3) gives the frequency equation:

$$\cos(\beta_n l)\cosh(\beta_n l) = -1 \tag{4}$$

Bring the solution in the frequency equation into Eq. (4) to get the resonance frequency of the single-layer micro-cantilever beam:

$$f_n = \left(\frac{(\beta_n l)^2}{2\pi}\right)\sqrt{\frac{\tilde{E} I}{\rho A l^4}} \tag{5}$$

The solution to the frequency is $\cos(\beta_n l)\cosh(\beta_n l) = -1$, the equation is $\beta_1 l = 1.875, \beta_2 l = 4.694, \beta_3 l = 7.855\ldots\ldots$

For a double-layer micro-cantilever structure composed of a fiber micro-cantilever with a thickness h_1 and a magnetostrictive film with a thickness of h_2, $\tilde{E} I$ is replaced by $\overline{\tilde{E} I}$, which is the equivalent bending stiffness, ρA is replaced by $\overline{\rho A}$, which is equivalent linear density,

$$\overline{\tilde{E} I} = \tilde{E}_s I_s + \tilde{E}_f I_f \tag{6}$$

$$\overline{\rho A} = \rho_s A_s + \rho_f A_f \tag{7}$$

Therefore, when the influence of gravity and other factors is ignored, the double-layered micro-cantilever structure is not affected by the magnetic field, and the magnetostrictive film does not produce a magnetostrictive effect. The resonance frequency when the double-layer cantilever is kept flat is:

$$f_n = \left(\frac{(\beta_n l)^2}{2\pi l^2}\right)\sqrt{\frac{\overline{\tilde{E} I}}{\overline{\rho A}}} \tag{8}$$

When the double-layer micro-cantilever structure is affected by the magnetic field, the magnetostrictive film produces a magnetostrictive effect, which causes the stress mismatch between the magnetostrictive film and the optical fiber micro-cantilever beam, which results in a double-layer micro-cantilever structure. Deflection (this article assumes that the double-layered micro-cantilever structure does not come into contact with the end face of the fiber after bending), and the double-layered micro-cantilever structure is shown in Fig. 2 after it is bent.

The axial strain ε_x of the double-layer micro-cantilever beam structure along the beam length direction is divided into two parts: tensile strain ε_s and bending strain ε_b, which is $\varepsilon_x = \varepsilon_s + \varepsilon_b$, The tensile strain ε_s reflects the change in the axial length

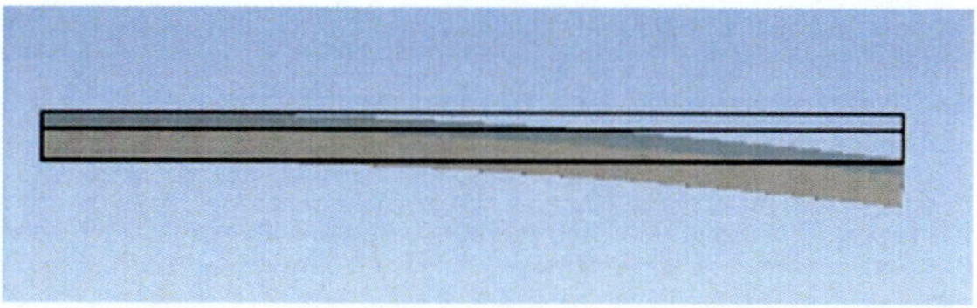

Fig. 2 The deflection of double-layer micro-cantilever beam structure

of the double-layer micro-cantilever structure, and the bending strain ε_b reflects the change in deflection. The resonance frequency of the double-layer micro-cantilever structure is mainly affected by the bending strain ε_b. Under the influence of the magnetic field of the double-layer micro-cantilever structure, when the deflection no longer changes, the stress between the magnetostrictive film and the optical fiber micro-cantilever reaches equilibrium $\sigma_s = \sigma_f$ [4], which is

$$\sigma = \sigma_s = \sigma_f = \frac{E_s}{1 - v_s^2}(\varepsilon_s + v_s \varepsilon_s) = \frac{E_f}{1 - v_f^2}\left[(\varepsilon_f - \lambda_s) + v_f\left(\varepsilon_f + \frac{\lambda_s}{2}\right)\right] \quad (9)$$

When the double-layer micro-cantilever structure is deflected, the expressions of the radius of curvature and the stress and equivalent bending stiffness between the magnetostrictive film and the optical fiber micro-cantilever are [5]:

$$R = \frac{\overline{\tilde{E}I}}{\int_s \sigma_s(z - z_c)dA_s + \int_f \sigma_f(z - z_c)dA_f} \quad (10)$$

z_c is the height of the neutral plane,

$$z_c = \frac{\left(\tilde{E}_s b\left(z_s^2\right) + \tilde{E}_f b\left(z_f^2 - z_s^2\right)\right)}{2\left(\tilde{E}_s b(z_s) + \tilde{E}_f b\left(z_f - z_s\right)\right)} \quad (11)$$

Assume that the double-layered micro-cantilever beam structure becomes a circular arc with a radius of curvature R as the radius after deflection [6]. A segment of the arch micro-elements is subjected to a force analysis [6],

$$\frac{\partial^6 v(\varphi, t)}{\partial \varphi^6} + 2\frac{\partial^4 v(\varphi, t)}{\partial \varphi^4} + \frac{\partial^2 v(\varphi, t)}{\partial \varphi^2}\frac{\overline{\rho A}}{\overline{\overline{\tilde{E}I}}}R^4\frac{\partial^4 v(\varphi, t)}{\partial \varphi^2 \partial t^2} = 0 \quad (12)$$

where φ is the angular coordinate in polar coordinates, t is time, and $v(\varphi, t)$ is the radial displacement component, let $(\varphi, t) = V(\varphi)\sin(\omega t + \theta)$, Substituting into Formula (12), we can get,

$$\frac{d^4 V(\varphi)}{d\varphi^4} + 2\frac{d^2 V(\varphi)}{d\varphi^2} + \left(1 - \chi^2\right)V(\varphi) = 0 \quad (13)$$

where $\chi^2 = \frac{\overline{\rho A}}{\overline{\tilde{E}I}} R^4 \omega^2$. In general, $\chi > 1$, with which as the boundary conditions, the general solution of Eq. (11) is obtained as,

$$\cos\left(\frac{\xi l}{R}\right)\cosh\left(\frac{\eta l}{R}\right) + \frac{1}{\xi\eta}\sin\left(\frac{\xi l}{R}\right)\sin\mathrm{h}\left(\frac{\eta l}{R}\right) + 1 = 0 \tag{14}$$

Let $\xi = \sqrt{\chi + 1}, \eta = \sqrt{\chi - 1}$, get

$$\cos\left(\frac{\sqrt{\chi+1}l}{\mathrm{R}}\right)\cosh\left(\frac{\sqrt{\chi-1}l}{\mathrm{R}}\right) + \frac{1}{\sqrt{\chi^2-1}}\sin\left(\frac{\sqrt{\chi+1}l}{\mathrm{R}}\right)\sin\mathrm{h}\left(\frac{\sqrt{\chi-1}l}{\mathrm{R}}\right) + 1 = 0 \tag{15}$$

By solving the above formula numerically, we can get the non-array definition of χ, χ_n (n = 1,2,3, …) It can be seen from the above formula that the resonance angular frequency ω corresponding to each χ value is the nth-order resonance angular frequency ω_n, Bringing the value of χ_n into the formula $\chi^2 = \frac{\overline{\rho A}}{\overline{\tilde{E}I}} R^4 \omega^2$ can obtain the resonance angular frequency ω_n. In this way, we can solve the resonance frequency F_n of each order of the double-layered micro-cantilever beam structure,

$$F_{nb} = \left(\frac{\chi_n}{2\pi R^2}\right)\sqrt{\frac{\overline{\tilde{E}I}}{\overline{\rho A}}} \tag{16}$$

Thus, the sensitivity of the double-layer micro-cantilever structure under the magnetic field can be obtained,

$$\mathrm{S} = \frac{\Delta f}{\Delta H} = \frac{F_{nb} - f_n}{\Delta H} = \frac{\left(\frac{\chi_n}{2\pi R^2} - \frac{(\beta_n l)^2}{2\pi l^2}\right)\sqrt{\frac{\overline{\tilde{E}I}}{\overline{\rho A}}}}{\Delta H} \tag{17}$$

3 Design and Optimization of Magnetic Field Sensor Structure

The magnetic field sensor is designed with an optical fiber micro cantilever on the end face of the optical fiber. The top surface of the optical micro cantilever is plated with a layer of magnetostrictive film, the magnetostrictive film forms a double-layer micro cantilever structure with the optical fiber micro-cantilever. The sensor structure is shown in Fig. 3.

Due to the limitation of the fiber size, in order to achieve the best sensitivity of magnetic field measurement, this paper needs to design the optimal size parameters of the double-layer micro-cantilever structure on the end face of the fiber. The main structures on the fiber end face are fixed ends and double-layer micro-cantilever

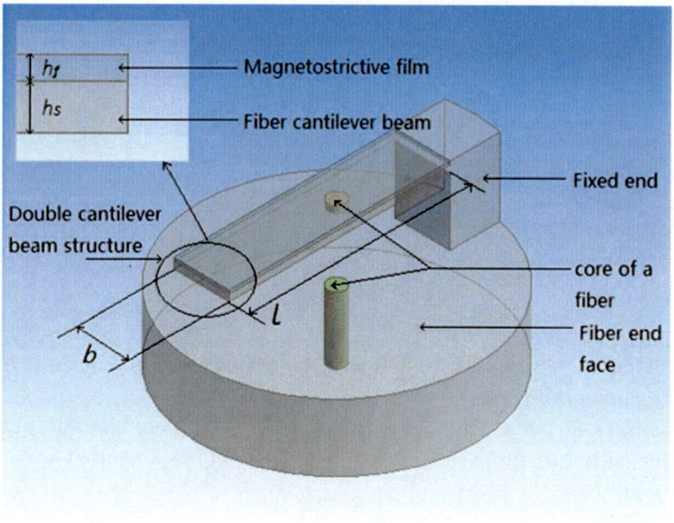

Fig. 3 The structure of the magnetic field sensor

beams. The sensitivity of the magnetic field sensor is mainly affected by the double-layer micro-cantilever structure. The sensitivity Formula of the magnetic field sensor is:

$$\mathrm{S} = \frac{\Delta f}{\Delta H} = \frac{(F_n - f_n)}{\Delta H} \tag{18}$$

Substituting the dimensional parameters of the double-layer micro-cantilever structure into Eq. (9) to simplify the first-order resonance frequency of the double-layer micro-cantilever structure without being affected by the magnetic field,

$$f_n = \frac{(1.875)^2}{2\pi l^2}\sqrt{\frac{\left(\frac{E_s h_s^3}{(1-v_s^2)} + \frac{E_f h_f^3}{(1-v_f^2)}\right)}{12(\rho_s h_s + \rho_f h_f)}} \tag{19}$$

According to Formula (17), among the structural parameters of the double-layer micro-cantilever beam, the length l of the beam, the thickness h_s of the optical fiber micro-cantilever beam, and the thickness h_f of the magnetostrictive film have an influence on the resonance frequency. Therefore, this paper uses ANSYS software to analyze the influence of the structural size parameters of the double-layer micro-cantilever on the sensitivity of the magnetic field sensor, and to obtain the structural size parameter with the highest sensitivity of the magnetic field sensor through simulation and optimization. This paper uses a single-mode fiber as a reference. The fiber diameter is about 125 μm and the core diameter is about 8 μm. The magnetic field is applied along the length of the cantilever beam and applied in a direction parallel

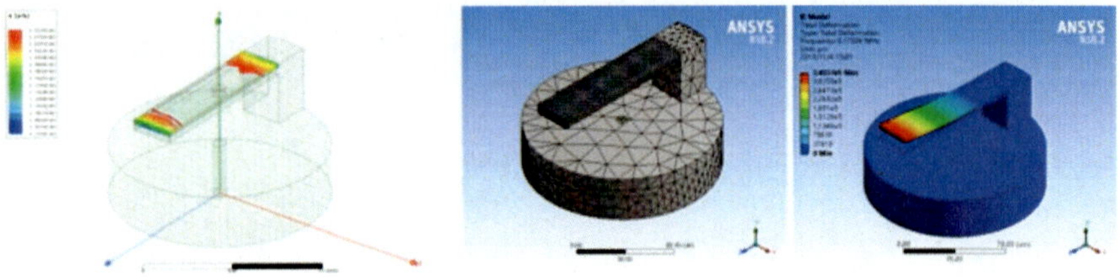

Fig. 4 The simulation flowchart in ANSYS

to the magnetic field. The grid uses a tetrahedral element to double-layer micro-cantilever. The beam structure is divided into 6000 elements, and the simulation flowchart is shown in Fig. 4.

First, the thickness h_s of the optical fiber micro cantilever and the thickness h_f of the magnetostrictive film are simulated and analyzed separately. Under the same magnetic field, the sizes of h_s and h_f are changed to obtain the sensitivity of the double-layer micro cantilever with different thicknesses, as shown in Fig. 5.

In Fig. 5, the abscissa is the ratio of the thickness of the magnetostrictive film to the thickness of the optical fiber micro-cantilever; the ordinate is the sensitivity of the double-layer micro-cantilever structure, and the unit is Hz/Gs. As can be seen from the figure, when the thickness of the optical fiber cantilever is 2 μm and the thickness of the magnetostrictive film is 3/5 of the thickness of the optical fiber cantilever under the same magnetic field, the sensitivity of the double-layer micro cantilever structure is the highest, which can reach 39,900 Hz/Gs.

At this time, the thickness of the fiber micro-cantilever and the magnetostrictive film were determined. Next, we simulated the influence of the structure length of the double-layer micro-cantilever on the sensitivity. Due to the fiber size limitation, the total length of the double-layer micro-cantilever structure and the fixed end must be equal to or less than 125 μm. In order to reduce manufacturing difficulties, the

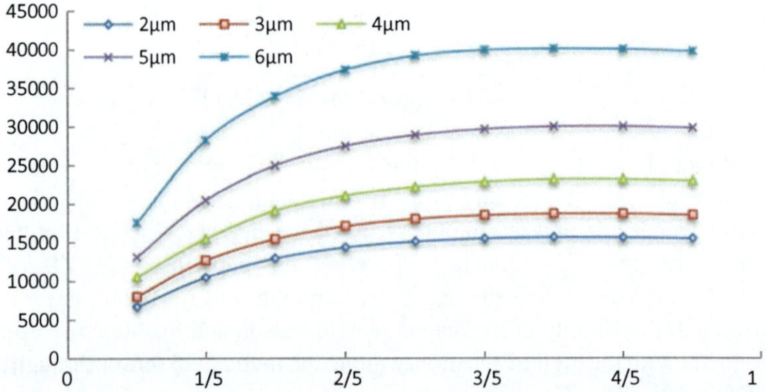

Fig. 5 The sensitivity of double-layer micro-cantilever structures with different thicknesses under magnetic field

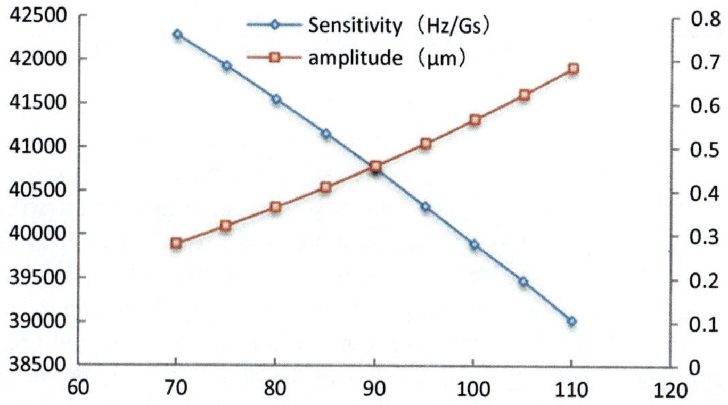

Fig. 6 The sensitivity and amplitude of double-layer micro-cantilever structures with different lengths

total length of the double-layer micro-cantilever structure and the fixed end is fixed at 125 μm. Only the length of the fixed-end is changed to achieve the change of the length of the double-layer micro-cantilever.

According to Formula (17), the larger the double-layer micro-cantilever structure is, the smaller the resonance frequency is. The larger the length of the double-layer micro-cantilever structure, the larger the resonance amplitude. The principle of detecting the vibration of a double-layer micro-cantilever structure is: detecting the change in the light intensity of the optical signal reflected by the structure back to the optical fiber. Therefore, the larger the resonance amplitude, the greater the change in light intensity, the higher the resolution of the frequency measurement, and the stronger the immunity to interference. Therefore, when optimizing the length of the double-layer micro-cantilever structure, we need to ensure that both the sensitivity and the amplitude are large. Under the same magnetic field, changing the size of l to obtain the sensitivity and amplitude of the double-layer micro-cantilever structure with different lengths, as shown in Fig. 6.

In Fig. 6, the abscissa is the length of the beam, the unit is micrometer; the ordinate is the sensitivity of the double-layer micro-cantilever beam, the unit is Hz/Gs. It can be seen from the figure that the sensitivity curve and amplitude curve of the double-layer micro-cantilever structure intersect at a beam length of 90 μm. Therefore, when the double-layer micro-cantilever structure length is 90 μm, compared with other double-layer micro-cantilever structures, the length has both a large sensitivity and a large amplitude. The sensitivity can reach 40,760 Hz/Gs and the amplitude can reach 0.46 μm.

4 Conclusion

This paper proposes a magnetostrictive fiber-optic micro-cantilever resonant magnetic field sensor, which is designed with an optical fiber micro cantilever on the end face of the optical fiber, and a layer of magnetostrictive film is plated on the optical fiber micro cantilever. The two constitute a double-layer micro cantilever structure. It uses frequency as the output signal and uses all-optical transmission without current interference. Therefore, this magnetic field sensor has the advantages of strong anti-interference ability and high sensitivity. Based on the important role of the double-layered micro-cantilever structure in the magnetic field sensor, the magnetic field-frequency characteristics of the double-layered micro-cantilever structure were simulated and optimized to obtain higher sensitivity. The optimal size after simulation optimization by ANSYS software is: when the double-layer micro-cantilever structure is 90 μm in length, the thickness of the optical fiber micro-cantilever is 2 μm, and the thickness of the magnetostrictive film is 3/5 of the thickness of the optical fiber micro-cantilever, the magnetic field sensor it can reach the maximum sensitivity of 40,760 Hz/Gs.

References

1. Shen F, Lu P, O'Shea SJ, Lee KH, Ng TY (2001) Thermal effects on coated resonant micro-cantilevers. Sens Actuators A Phys 95(1):17–23
2. Bigaud D, Szostkiewicz C, Hamelin P (2003) Tearing analysis for textile reinforced soft composites under mono-axial and bi-axial tensile stresses. Compos Struct 62(2):129–137
3. Forster B, Mollaert M (2004) European design guide for tensile surface structures. Brussel: Tensi-Net 293–322
4. Ripka P (2003) Advances in fluxgate sensors. Sens Actuators A 106(1):8–14
5. Du T, de Lacheisserie E (1993) Magnetostriction: theory and applications of magnetoelasticity. CRC Press, Boca Raton, FL
6. SeshiaAA, PalaniapanM,RoessigTA, HoweRT, GoochRW,SchimertTR,MontagueS (2002) A vacuum packaged surface micromachined resonant accelerometer. J Microelectromech Syst11(6):784–793

Theoretical Model and Optimum Design of Optical Fiber Micro-cantilever Beam Magnetic Field Sensor

Cheng Xu, Yueming Liu, and Yuchan Liu

Abstract A new structure of fiber micro-cantilever beam magnetic field sensor probe is proposed. A micro-cantilever beam was fabricated on the end face of the fiber, and the surface of the fiber micro-cantilever beam was coated with terfenol-D, a magnetostrictive material that can be flexed according to the magnetic field. Due to the optical path is changed by the Fabry–Perot cavity deflection, the Fabry–Perot cavity output light intensity is changed, and the magnetic field can be detected by the change in light intensity. The advantages of the sensor probe are structural simplicity, high magnetic field measurement sensitivity, and All-optical transmission without electrical excitation interference. Based on the minimum energy theory, the formula of deflection with magnetic field was derived, the calculated average sensitivity of the probe is 1.3 nm/nT, the minimum detectable magnetic field is 0.1nT. The structure of the sensing probe has optimized by ANSYS finite element analysis software.The optimized sizes of the probe were as follows: the length of optical fiber cantilever beam is 104 μm, The width of optical fiber cantilever beam is 22 μm, the length of the fixed support is 20 μm, the thickness of the micro-cantilever optical fiber layer is 8 μm, the thickness of the chrome layer is 0.04 μm, the thickness of the magnetostrictive film is 4 μm, and the initial cavity length of the F-P cavity is 16 μm.

Keywords Optical Fiber Micro-cantilever · Fabry–perot (f-p) · Mgnetostrictive Film · ANSYS · Deflection

1 Introduction

Magnetic navigation, seismic monitoring, biomedical, geological exploration, electronic equipment testing and military guidance [1, 2]. Conventional magnetic field

C. Xu · Y. Liu (✉) · Y. Liu
China Jiliang University, HangZhou, China
e-mail: 07a0402054@cjlu.edu.cn

Y. Peng and X. Dong (eds.), *Proceedings of 2019 International Conference on Optoelectronics and Measurement*, Lecture Notes in Electrical Engineering 726,
https://doi.org/10.1007/978-981-33-4110-4_30

sensors are usually of the type of electrical signals such as Hall magnetic sensors, magnetoresistive sensors and fluxgate sensors [3–5]. Such magnetoelectric sensors usually use active metal wires to provide current excitation, also, the measured magnetic field produces additional interference, which is bulky based on the volume, result in the difficulty to miniaturize the sensors.

In recent years, many domestic and foreign scholars specializing in the field of optical fiber magnetic field sensor performed research studies using MOEMS, such as a fiber-optic magneto-optical crystal sensor based on a Faraday magnetron effect [6], a fiber-optic Michelson interference, a Fabry–Perot (FP) interference sensor, or a Mach–Zehnder (MZ) interference sensor coated with a fiber magnetostrictive material [7], Fiber Bragg Grating (FBG) sensors coated with magnetically sensitive materials [8], and various optical fiber magnetic field micro-sensors with magnetic fluid and fiber photonic crystal structures [9]. MOEMS technology has advantages over traditional mechanical technology, especially on account of high sensitivity and compact size. It can overcome the shortcoming of current electrical magnetic field sensor from the structure and principle.

In this paper, a novel structure of optical fiber Micro-cantilever magnetic field sensor structure is designed. The magnetic micro-cantilever beam was fabricated on the end face of the fiber by MOEMS technology to form the sensor, based on the principle of the optical fiber Fabry–Perot (F-P) interference, and magnetizes the magnetostrictive film on the surface of the fiber micro-cantilever beam to achieve the all-optical fiber magnetic sensor detection without current excitation. This fiber-optic magnetic sensor has the advantages of optical fiber integration, all-optical transmission, miniaturization, high sensitivity and strong anti-interference performance [10, 11]. Under the premise of optimized design, the detection sensitivity is 1.3 nm/nT, and the minimum detectable magnetic field is expected to increase to 0.1nT. New structure probe can be applied to the narrow and complex space that is difficult to reach with traditional magnetic sensors, and is used for the detection of weak magnetic fields.

2 Design of Sensor

The optical fiber cantilever beam proposed in this paper is located at the end face of the optical fiber, the structure of the magnetic field sensor probe has been shown in Fig. 1. In the middle of the sensing probe is a fiber core. One end of the micro-cantilever is fixed, one end is free, and an F-P interference cavity is formed between the surface of the micro cantilever beam and the end face of the fiber.

The optical fiber micro-cantilever beam is sequentially plated with a chrome layer and a giant magnetostrictive film. The chrome layer serves as a buffer layer, which can bond the giant magnetostrictive film to the optical fiber cantilever beam, and increase the reflectivity and reduce the optical wave loss.

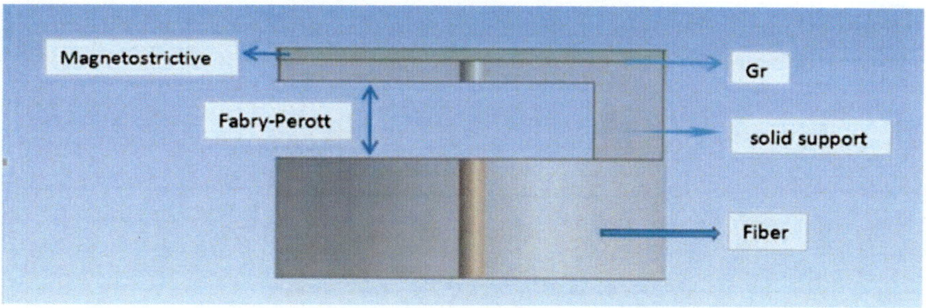

Fig. 1 The structure of micro-cantilever beam magnetic field sensor

Under the action of the magnetic field, the magnetostrictive layer drives the magnetic micro-cantilever beam to be deflection. By detecting the magnetostrictive deflection of the magnetic micro-cantilever beam, the strength of the magnetic field can be measured. Technically, the deflectionof the optical fiber micro-cantilever beam can be detected by Fabry–Perot cavity output light intensity. Fabry–Perot cavity is formed between the optical fiber end face and the micro-cantilever beam, the theoretical detection capability of the Fabry–Perot interference cavity can reach sub-nanometer level, and the minimum detectable magnetic field is 0.1nT based on sensitivity.

3 Theoretical Model of Magnetic Field Measurement

The total energy of the cantilever fiber substrate is elastic energy, and the elastic energy density Formula is [12],

$$F_s = \frac{1}{2}\frac{Y_s}{(1+v_s)(1-2v_s)}\left[(1-v_s)\left(\varepsilon_1^2+\varepsilon_2^2+\varepsilon_3^2\right)+2v_s(\varepsilon_1\varepsilon_2+\varepsilon_2\varepsilon_3+\varepsilon_1\varepsilon_3)\right] \tag{1}$$

where v_s is the Poisson's ratio of the cantilever beam optical fiber layer, Y_s is the elastic modulus of the cantilever beam optical fiber layer, ε_1,ε_2,ε_3 are Three tiny quantities on the X, Y, and Z axes, respectively. Under plane stress conditions, the free plate with unconstrained surface has zero stress in the Z-axis direction.

$$\frac{\partial F}{\partial \varepsilon_3} \tag{2}$$

$$\varepsilon_3 = -\frac{2}{D(1-V_s)}(\varepsilon_1+\varepsilon_2) \tag{3}$$

Substituting (3) into (1) can obtain the energy density expression of the micro-cantilever optical fiber layer represented by ε_1 and ε_2.

$$F_s(\varepsilon_1, \varepsilon_2) = \frac{Y_s}{2(1 - v_s^2)}\left(\varepsilon_1^2 + \varepsilon_2^2 + 2v_s\varepsilon_1\varepsilon_2\right) \tag{4}$$

The total energy of the micro-cantilever optical fiber layer is,

$$E_S^T = \iiint\limits_S F^S(\varepsilon_1, \varepsilon_2)dV = \int_0^L dx \int_0^w dy \int_{-d_s}^0 F^S(\varepsilon_1, \varepsilon_2)dz = \frac{V_sY_s}{2(1 - v_b^2)} \times \left\{\delta_1^2 - \delta_1\gamma_1 d_s + \frac{\gamma_1^2 d_s^2}{3} + \delta_2^2 - \delta_2\gamma_2 d_s + \frac{\gamma_2^2 d_s^2}{3} + 2v_s\left[\delta_1\delta_2 - \frac{d_s}{2}(\delta_1\gamma_2 + \delta_2\gamma_1) + \frac{\gamma_1\gamma_2 d_s^2}{3}\right]\right\} \tag{5}$$

where γ_1 and γ_2 are the curvature radius of the Micro-cantilever optical fiber layer along the length and width directions, respectively, V_s and d_s are the volume and height of the micro-cantilever optical fiber layer, respectively.

The energy of the magnetostrictive layer consists of two parts: magnetoelastic energy and elastic energy,

$$F_f = F_m + F_e \tag{6}$$

The energy density of magnetoelastic energy is as follows

$$F_m = -A\varepsilon_1 - B\varepsilon_2 - C\varepsilon_3 \tag{7}$$

A, B, and C are undetermined coefficients, corresponding to the dimension of stress, and the elastic energy density of the magnetostrictive layer is as follows,

$$F_e = \frac{1}{2}\frac{Y_f}{(1 + v_f)(1 - 2v_f)}\left[(1 - v_f)\left(\varepsilon_1^2 + \varepsilon_2^2 + \varepsilon_3^2\right) + 2v_f(\varepsilon_1\varepsilon_2 + \varepsilon_2\varepsilon_3 + \varepsilon_1\varepsilon_3)\right] \tag{8}$$

v_f and Y_f are the Poisson's ratio and Young's modulus of the magnetostrictive material, respectively. According to the principle of energy minimization, the deformation of the cantilever beam under the action of an external magnetic field should minimize the total energy, A, B, C can be expressed as follows,

$$A = \frac{Y_f}{(1 + v_f)(1 - 2v_f)}\left[\varepsilon_1^0(1 - v_f) + v_f\left(\varepsilon_2^0 + \varepsilon_3^0\right)\right] \tag{9}$$

$$B = \frac{Y_f}{(1+v_f)(1-2v_f)}\left[\varepsilon_2^0(1-v_f)+v_f(\varepsilon_1^0+\varepsilon_3^0)\right] \tag{10}$$

$$C = \frac{Y_f}{(1+v_f)(1-2v_f)}\left[\varepsilon_3^0(1-v_f)+v_f(\varepsilon_2^0+\varepsilon_1^0)\right] \tag{11}$$

$\varepsilon_1^0,\varepsilon_2^0,\varepsilon_3^0$ are the strain when the energy is minimum. if the magnetostrictive layer is an isotropic material and the magnetostrictive effect under the action of an external magnetic field is anisotropic, the strain of the micro-cantilever due to magnetostriction is as follows [13],

$$\varepsilon_1^0 = -2\varepsilon_2^0 = -2\varepsilon_3^0 = \lambda_f \tag{12}$$

where λ_f is the magnetostriction coefficient, the energy density phase of the magnetostrictive layer is calculated as follows,

$$F_f(\varepsilon_1,\varepsilon_2) = \frac{Y_f\lambda_f}{2\left(1-v_f^2\right)}\left[\varepsilon_1(2-v_f)-\varepsilon_2(1-2v_f)\right] + \frac{Y_f}{2\left(1-v_f^2\right)}\left(\varepsilon_1^2+\varepsilon_2^2+2v_f\varepsilon_1\varepsilon_2\right)\frac{Y_f\lambda_f^2(1-2v_f)}{8\left(1-v_f^2\right)} \tag{13}$$

By integrating the above Formula, the total energy of the magnetostrictive layer can be obtained as follows [14],

$$\begin{aligned} E_f^T &= \iiint\limits_f F_f(\varepsilon_1\varepsilon_2)dV = \int_0^L dx\int_0^\omega dy\int_0^{d_f} dx F_f(\varepsilon_1\varepsilon_2)dz \\ &= -\frac{V_fY_f\lambda_f}{2\left(1-v_f^2\right)}\left[(2-v_f)\left(\delta_1+\frac{\gamma_1 d_f}{2}\right)-(1-2v_f)\left(\delta_2+\frac{\gamma_2 d_f}{2}\right)\right] \\ &+ \frac{V_fY_f}{2\left(1-v_f^2\right)}\left\{\delta_1^2+\delta_1\gamma_1 d_f+\frac{\gamma_1^2 d_f^2}{3}+\delta_2^2-\delta_2\gamma_2 d_f+\frac{\gamma_2^2 d_f^2}{3}\right. \\ &\left.+2v_f\left[\delta_1\delta_2+\frac{d_f}{2}(\delta_1\gamma_2+\delta_2\gamma_1)+\frac{\gamma_1\gamma_2 d_f^2}{3}\right]\right\} \end{aligned} \tag{14}$$

where $V_f = Lwd_f$ the Volume of magnetostrictive layer, $V_s = Lwd_s$ is the volume of the cantilever beam optical fiber layer, δ_1, δ_2 are the amount related to the neutral surface, respectively. Adopting the principle of minimum potential energy for the entire micro-cantilever beam, the deflection can be expressed as follows [15],

$$\Delta = \frac{1}{2}L^2(\gamma_1 - \gamma_2) = \frac{9}{2}L^2\frac{qH(d_f + d_s)}{d_f d_s}$$
$$\times\left[\left(\frac{d_f}{d_s}\right)^2\frac{Y_f/(1+\upsilon_f)}{Y_s/(1+\upsilon_s)} + 4\left(\frac{d_f}{d_s}\right) + 6\right.$$
$$\left. +4\left(\frac{d_s}{d_f}\right) + \left(\frac{d_s}{d_f}\right)^2\frac{Y_s/(1+\upsilon_s)}{Y_f/(1+\upsilon_f)}\right]^{-1} \quad (15)$$

4 Structural Optimization of Sensor

4.1 Selection of Initial Cavity Length of F-P Cavity

A monochromatic light source with a wavelength of λ is used, and the light intensity of the photo detector is as following [16],

$$I_R = 2R[1 - cos(4\pi L/\lambda)]\left[1 + R^2 - 2Rcos(4\pi L/\lambda)^{-1}\right]I_0 \quad (16)$$

where R is the reflectivity of cavity section, L is the length of cavity, λ is the wavelength of the incident light wave, and I_R is a uni-variate function of the cavity length L (Fig. 2).

When the cavity length $L = (2m + 1)\lambda/4$, the sensor outputs a maximum value I_R; when the cavity length $L = 2m\lambda/4$, the sensor outputs a minimum value I_R. I_R is the multi-valued function of L. The initial cavity length of the sensor is limited to less than $\lambda/2$, as the Δ region shown in Fig. 3, and the light intensity is a single-valued function of L. According to the structure size of the sensor, in order to make the

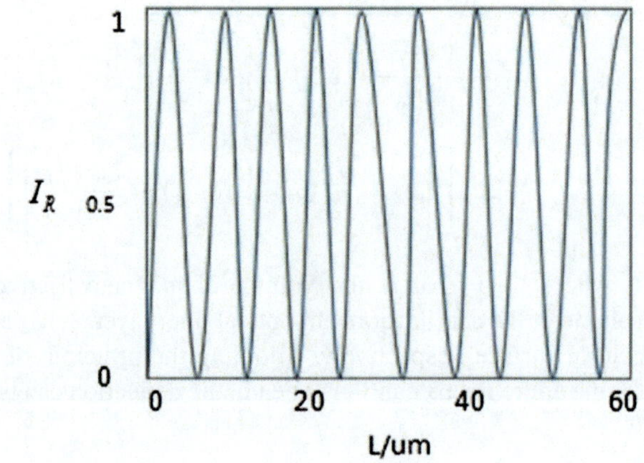

Fig. 2 Output light normalization intensity of overall periodicity

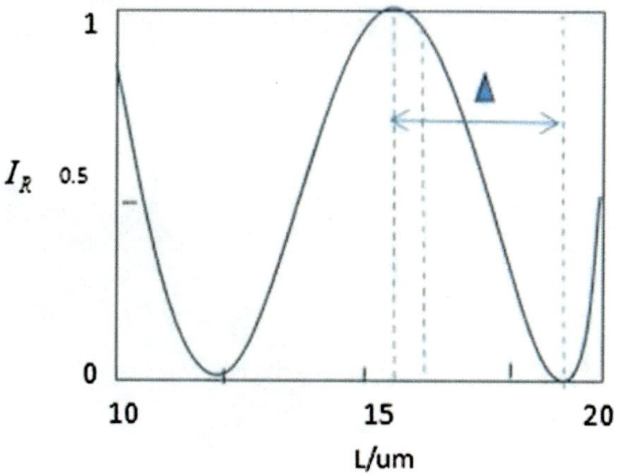

Fig. 3 Detail of optical fiber F-P output under monochromatic light conditions

sensing probe more sensitive, the area with the larger slope in the Fig. 3 is selected, and the initial cavity length is determined to be 16 μm.

4.2 *Selection of Magnetostrictive Materials*

At present, there are three main types of commonly used magnetostrictive materials: (1) magnetostrictive materials of metals and alloys, such as: Ni metal, Fe–Ni alloy, Fe-Co-Cr alloy. (2) Ferrite magnetostrictive materials, such as Ni-Co ferrite materials, etc. (3) Rare earth intermetallic compound magnetostrictive materials. The performance of several magnetostrictive materials is shown in Table 1

As can be seen from Table.1, the magnetostriction coefficient of Terfenol-D is several hundred times of ordinary magnetostrictive materials. It has high sensitivity, fast reaction speed and high reliability. Therefore, in this study, the Terfernol-D target was selected, and the magnetostrictive film was deposit on the surface of the fiber-end, which can make the probe miniaturized and greatly improve the performance.

Table 1 Table of magnetostrictive material performance comparison

material	Fe-Co-Cr alloy Cr:0.4–0.5 Co:34.5- 35.5	Ni-Co Ferrite Ni:1.0 Co:0.012	Terfenol-D Tb0.27 Dy0.73 Fe
Saturation magnetostricti on coefficient(ppm)	40	-27	1500–200
Young's modulus (1010 Pa)	20.6	17.0	2.65
Curie temperature (°C)	1115	590	387

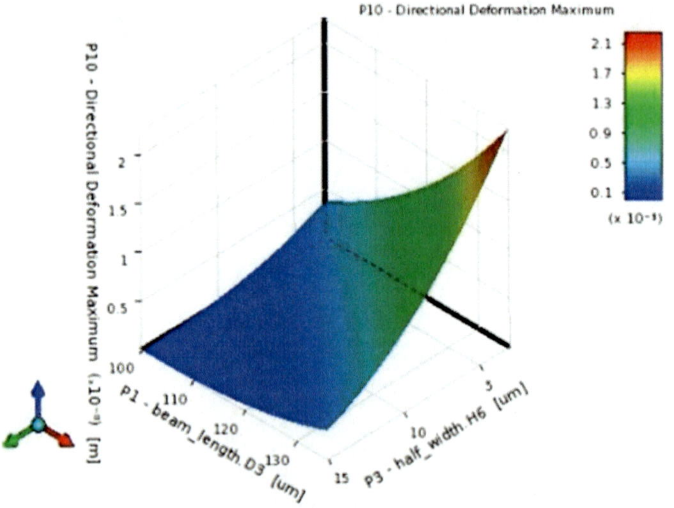

Fig. 4 The diagram of the relationship between the length, width, and deflection of a beam

4.3 Structural Optimization of Micro-Cantilever

The finite element calculation was done in ANSYS Workbench. The geometry model is built in the geometry module of the Workbench platform. For the parameters that need to be optimized, the upper and lower limits of the change are specified for each parameter. The material parameters calculated by the electromagnetic field are specified in the maxwell 3D module, and the material parameters calculated by the structure are defined in the engineer data of the static structural module. A tangential H field load is applied on four faces parallel to the direction of the uniform magnetic field. The response of the two parameters to the deflection is shown in Figs. 4 and 5

In the theoretical analysis model of fiber magnetostrictive micro- cantilever deflection, the ANSYS is used to optimize the six structural parameters to maximize the deflection. The parameters include the length of the optical fiber beam (including the length of the fixed support), the width of the micro-cantilever beam (the 1/2 model), the length of the fixed support, the thickness of the micro- cantilever optical fiber layer, the chromium layer thickness, and magnetostrictive film thickness. The optimization results are shown as follow, (Fig. 6)

The structural optimization of the fiber-optic magnetic field probe is realized by ANSYS. These three columns are recommended for optimization, the first column is selected as the optimal structure: the length of fiber micro-cantilever beam is 104 μm, the width of cantilever beam is 22 μm, the length of fixed-support is 20 μm. The thickness of the micro-cantilever optical fiber layer is 8 μm, the thickness of the chrome layer is 0.04 μm, and the thickness of the magnetostrictive film is 4 μm (Fig. 7).

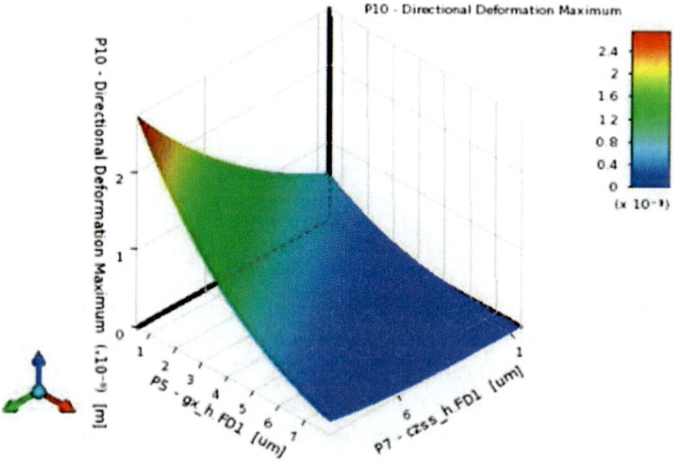

Fig. 5 The relationship between the fiber layer thickness, the magnetostrictive layer thickness and the deflection

15		Candidate Point 1	Candidate Point 2	Candidate Point 3
16	P4 - liang.c (um)	★★★ 123.68	★★ 123.36	– 117.6
17	P2 - liang.k (um)	★★★ 10.995	★★★ 11.651	★★★ 11.61
18	P3 - gzd.c (um)	★★★ 20.315	★★ 22.414	★★★ 20.256
19	P5 - liang.h (um)	★★ 8.7601	★★★ 7.9201	★★ 8.5345
20	P6 - cr.h (um)	★ 0.043895	★★ 0.040833	★ 0.04883
21	P7 - czss.h (um)	★★★ 4.1396	★★ 4.4371	★★ 4.7256
22	P9 - Total Deformation Maximum (um)	★ 3.8014E-09	– 4.4788E-09	★★ 2.0766E-09
23	P8 - Deformation Probe Maximum Z Axis (um)	★★★ -6.7665E-11	★★ 2.7204E-09	★★★ 1.3475E-10

Fig. 6 The relationship between the fiber layer thickness, the magnetostrictive layer thickness and the deflection

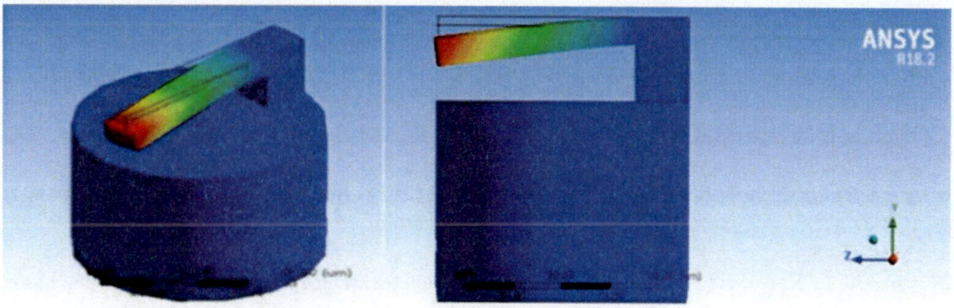

Fig. 7 The deflection of the optical fiber micro-cantilever beam

The saturation magnetostriction coefficient $\lambda_f = qh = 2000$, $Y_f = 7.31E + 10$, $V_f = 0.17$,$Y_S = 3E + 10$, $V_S = 0.3$ and the optimized size were substituted into Eq. (16), and it was concluded that the deflection was 2um and the sensitivity was 1.3 nm/nT.

5 Conclusion

A novel structure of optical fiber cantilever magnetic-filed sensor probe structure is proposed. The optical fiber integrated micro-machining process is adopted, and the Terfenol-D film is plated on the surface of the fiber micro-cantilever beam. The deflection can be detected by the F-P interference cavity and realizes the all-optical fiber transmission without current excitation. The optimal dimensions of the probe are as follows: the length of optical fiber cantilever beam is 104 μm, The width of fiber cantilever beam is 22 μm, the fixed support length is 20 μm, the thickness of the micro-cantilever optical fiber layer is 8 μm, the thickness of the chrome layer is 0.04 μm, the thickness of the magnetostrictive film is 4 μm, and the initial cavity length of the F-P cavity is 16 μm. The optimized magnetic field sensor has an average sensitivity of 1.3 nm/nT and a minimum detectable magnetic field of 0.1nT. It can be used in small and complex spaces that are difficult to reach with traditional magnetic sensors. Magnetic field detection is a high performance fiber optic miniature magnetic sensor probe with a new structure and sensing mechanism.

References

1. Velasco-Quesada G,Roman-Lumbreras M,Conesa-Roca A (2011) Design of a low-consumption fluxgate transducer for high-current measurement application.IEEE Sensor J 11(2):280-287
2. VyhnanekJ,JanosekM,RipkaP (2012)AMR gradiometer for minedetection.Sens Actuators A-Phys 186(SI):100–104
3. Ripka P (2003) Advances in fluxgate sensors. Sens Actuators A-Phys 106(1):8–14
4. Hans H, Johann H, Günther S, Rupert C, Michael J, Karl R (2000) Anisotropic magnetoresistance effect field sensors. J Magn Magn Mater 215:788–879
5. Paun MA,SalleseJM,Kayal M (2012) Offset and drift analysis of the hall effect sensor. The geometrical parameters influence. Digest J NanomaterBiostruct7(3):883–891
6. Yin J, Ruan S, Liu T, Jiang J, Wang S, Wei H, Yan P (2017) All-fiber-optic vector magnetometer based on nano-magnetic fluids filled double-clad photonic crystal fiber. Sens Actuators B: Chem 238:518–524
7. Zhao Y, Li X, Cai L (2015) A highly sensitive Mach-Zehnder interferometric refractive index sensor based on core-offset single mode fiber. Sens Actuators A 223:119–124
8. Chen Y, Han Q, Yan W, Yao Y, Liu T (2016) Magnetic-fluid-coated photonic crystal fiber and FBG for magnetic field and temperature sensing. IEEE Photon Technol Lett 28(23):2665–2668
9. Lu K, Yang H, Lim KS, Ahmad H, Zhang P (2018) Effect of two annealing processes on the thermal regeneration of fiber Bragg gratings in hydrogenated standard optical fibers. Appl Opt 57(24):6971–6975

10. Violakis G, Korakas N, Pissadakis S (2018) Differential loss magnetic field sensor using a ferrofluid encapsulated D-shaped optical fiber. Opt Lett 43(1):142–145
11. Du Trémolet de LacheisserieE, PeuzinJC (1994) Magnetostriction and internal stresses in thin films: the cantilever method revisited. J Magnet Magnetic Mater136(1–2):189–196
12. Kazushi I, Chikako Y (2008) Cantilevered actuator using magnetostrictive thin film. J Magn Magn Mater 320(20):2481–2484
13. Jiles DC (1995) Theory of the magnetomechanical effect. J Phys D Appl Phys 28(8):1537–1546
14. Blackburn JF, Vopsaroiu M, Cain MG (2008) Verified finite element simulation of multiferroic structures: Solutions for conducting and insulating systems. J Appl Phys 104(7):074104
15. Li J, Wen Y, Li P, Yang J (2017) Modeling of magnetoelectric effects in magnetostrictive/piezoelectric laminated composites using the energy method. IEEE Trans Magn 53(9):2500406
16. Shi F, Luo Y, Che J, Ren Z, Peng B (2018) Optical fiber F-P magnetic field sensor based on magnetostrictive effect of magnetic fluid. Opt Fiber Technol 43:35–40

Recent Development in Electro-thermal Modeling and Simulation of OLEDs

Changfeng Gu, Qinyong Dai, Juanjuan Zhou, Xinyu Song, Zhuoli Zhou, Chengyu Lu, and Yingquan Peng

Abstract Compared with traditional LEDs, OLEDs have a lower luminous efficiency, so the surface must be much larger to provide a light output equivalent to the total luminous flux. Brightness uniformity is a challenge for equipment engineers. High brightness will increase power consumption, causing Joule heat to self-heat. An increase in temperature may cause the performance of the OLED to decrease, and also cause the lateral brightness to become very uneven. Therefore quantitatively understand the thermal environment of multilayer composite devices under high current, and use this as a guide to mitigate the impact of heating by optimizing device and system design becomes very important. In this paper, recent progress regarding to the research on electro-thermal modeling of OLEDs is reviewed, with respect to the fundamental understanding of modeling of eletro-thermal processes. The current state of the simulation research is summarized and simulation methods is described.

Keywords Electro thermal model · Simulation · OLED

1 Introduction

Modern organic light-emitting device (OLED) display technology has the advantages of high contrast, low energy consumption, fast response time and large viewing angle [1]. OLEDs have strong temperature dependence and have nonlinear electrical characteristics. On one hand, partial Joule heating reduces efficiency, service life, and brightness uniformity; on the other hand in a natural convection environment, since the temperature difference of the device will reach 20–30 °C, its brightness will vary by as much as 30–40%. It is reported that a 10 °C increase in temperature will increase

C. Gu · Q. Dai · J. Zhou · X. Song · Z. Zhou · C. Lu · Y. Peng (✉)
Institute of Microelectronics, College of Optical and Electronic Technology, China Jiliang University, Hangzhou, China
e-mail: yqpeng@cjlu.edu.cn

Y. Peng and X. Dong (eds.), *Proceedings of 2019 International Conference on Optoelectronics and Measurement*, Lecture Notes in Electrical Engineering 726,
https://doi.org/10.1007/978-981-33-4110-4_31

the conductivity of the junction by 40–50% [2]. For lighting applications, uniform emissions are required, which is closely related to spatial temperature distribution [3–7]. Therefore quantitatively understand the thermal environment of multilayer composite devices under high current, and use this as a guide to mitigate the impact of heating by optimizing device and system design becomes very important. Therefore greater attention has been focused on the thermal aspects of OLEDs in simulation [8]. This review describes the models and methods developed in recent years for elctro-thermal simulation of OLEDs.

2 Electrothermal Simulation

Gielen et al. developed a finite element-based simulation model, in which two-dimensional (2D) model and three-dimensional (3D) model can be chose in different situations, to study the electrical-thermo-mechanical interaction in OLEDs [9]. Here, 2D model was suitable for studying several choices regarding layers and basic layout, while 3D model was needed for design details such as the bus bar layout. The simplified 2D model was used to explore the solution strategy and to study two design variants.

As shown in Fig. 1, in the variant A, the bus bar is in the Barrier layer; in the variant B, it is embedded in the LEP layer and the ITO layer. In the two variants, the temperature distribution along the LEP layer and the stress distribution along the layer are similar. Variant A consumes 15–20% more electricity than variant B. Furthermore, the stresses in the variants C and D due to electro-thermo-mechanical coupling were shown in Fig. 3, in which these stresses were in a similar range to the stresses determined in the 2D model and variants D exhibited a higher stresses value. All of this suggested that different design choices will affect different quality aspects. (Fig. 2).

Pohl et al. created an electro-thermal model for the OLED samples provided by their project partner, in which the algorithm in the simulation tool, SUNRED,

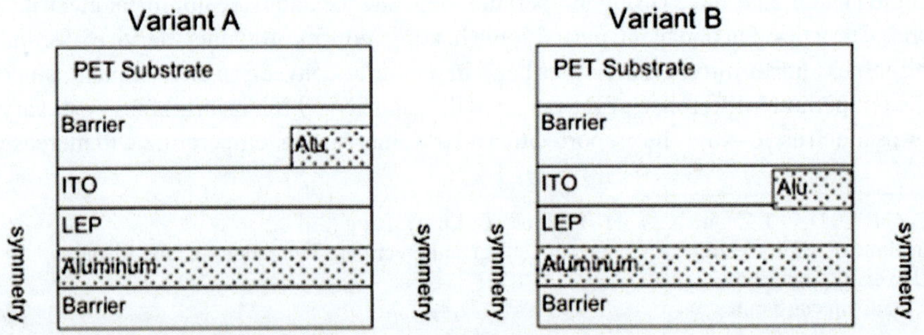

Fig. 1 Two-dimensional model diagram of OLEDs. Every layer is 200 nm thick. The PET substrate is a boundary condition instead of a layer in the model [9]

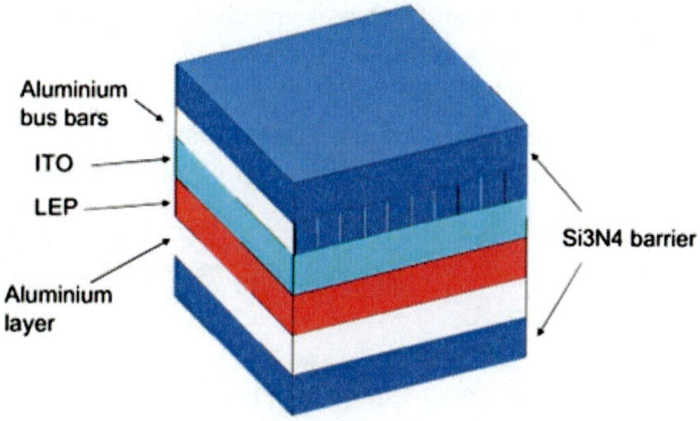

Fig. 2 Three-dimensional OLED model (z-axis size is enlarged relative to x and y axes) [9]

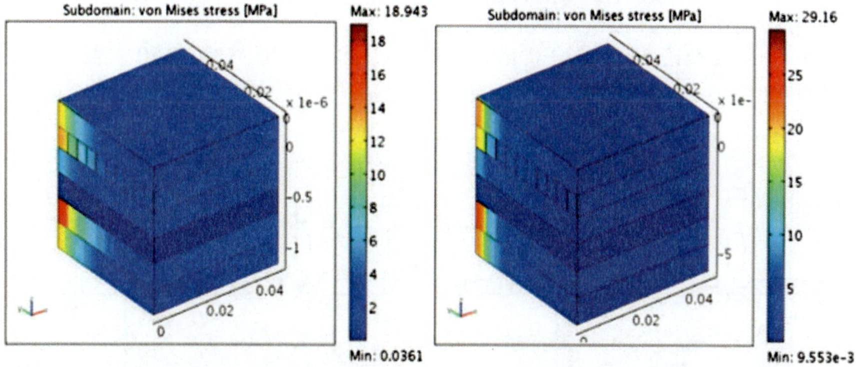

Fig. 3 Stress distribution in variant C (left) and variant D (right) [9]

had been extended by using a nonlinear semiconductor junction model [10]. Therefore, this extension can be used to handle any nonlinear function including highly nonlinear OLED electronic characteristics and the nonlinear temperature dependence. The organic light emitting diode with thin-film multi-layer was shown in Fig. 4, which need distributed electrical simulation to predict the voltage drop over

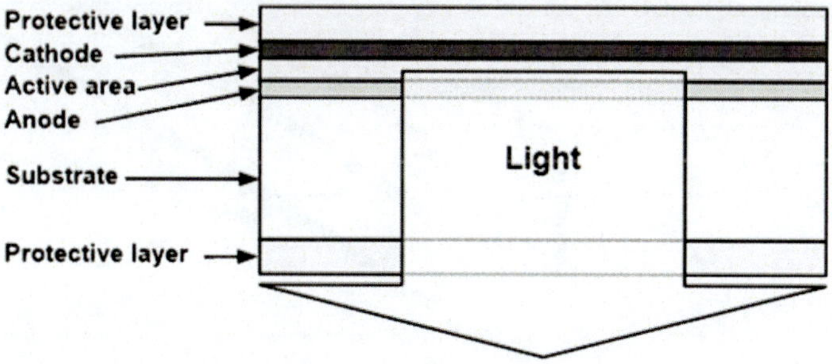

Fig. 4 The layer structure of the OLED device [10]

the large surface, and thermal simulation to examine the temperature distribution over the surface or inside the layer structure.

The original SUNRED model was modified as shown in Fig. 5 due to the semiconductor junction is a thin nonlinear surface between two electrically nearly linear materials. And the result of the junction modeled by power function was as follows:

$$G_j = \frac{I}{\sqrt[m]{\frac{A_{full}}{bA_{cellside}}}} \quad (1)$$

where m and b are calculated befor, I is the current of G_j, A_{full} is the area of the full semiconductor junction (all connecting cells), $A_{cellside}$ is the area of the junction side of the actual cell. Finally, the maximal relative error between two steps decreased under 0.0001% via 53 iterated computation.

For comparison, 60 nm LEP thickness and 10 °C, 30 °C and 50 °C ambient temperatures were selected. As described in Fig. 6, the resulting curves including

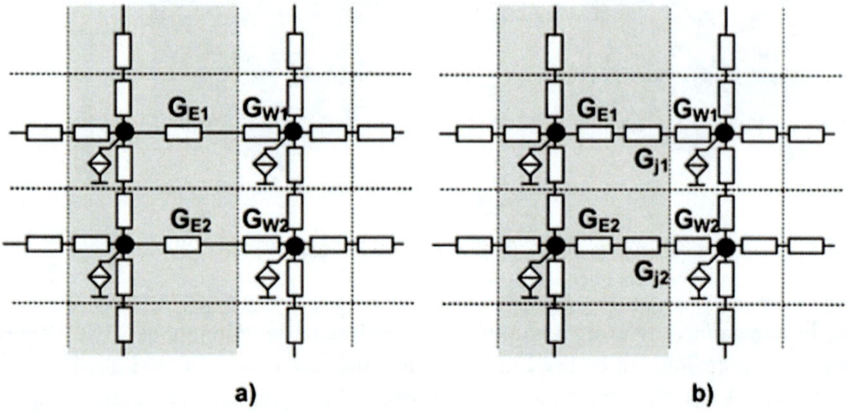

Fig. 5 The SUNRED model for normal materials **a** and semiconductor junctions **b** [10]

Fig. 6 Measurement, model and simulation results compared, 60 nm LEP [10]

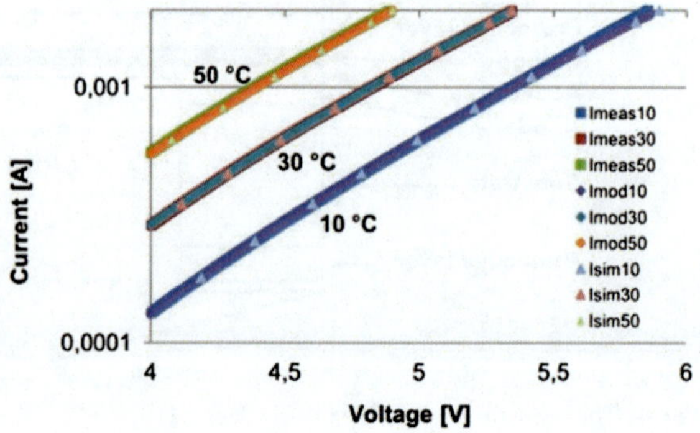

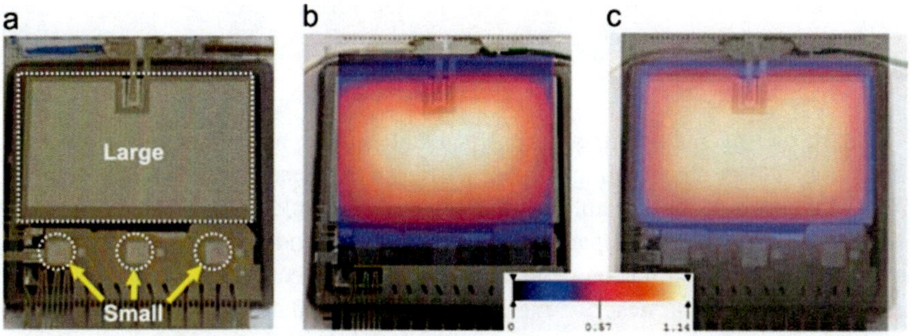

Fig. 7 **a** A sample device with the investigated OLEDs. **b** Measured and **c** simulated surface temperature distribution at a heating power level of 565 mW, with temperature scale [11]

three measured curves, three model solution result curves and three simulation result curves were mutually overlapping, which confirmed that the model and simulation are correct.

Organic semiconductor materials used in OLEDs have highly nonlinear electrical characteristics and are closely related to temperature. La´szlo´ Pohl et al. gave the latest SUNRED electrothermal field solver algorithm, which is suitable for coupled nonlinear electrothermal (as shown in Fig. 7) [11].

The latest version of the SUNRED field reducer algorithm by Ernő Kollár et al. can deal with these exchanged nonlinear photoelectric thermal problems and solve the special needs of OLEDs in distributed electric thermal field simulation [12]. They modeled the OLEDs samples from the Fast2Light project and compared the simulation results with the measured data, proving that the new features of the algorithm are suitable for the development of OLEDs. The resulting model equations for the sample OLEDs are: (Fig. 8)

$$I_{OLED,60nm}(U,T) = \left(0.397 \times T^2 - 5.47 \times T + 94.2\right) \times 10^{-10} \times U^{1.4\times10^{-4}\times T^2 - 2.55\times10^{-2}\times T + 7.28} \tag{2}$$

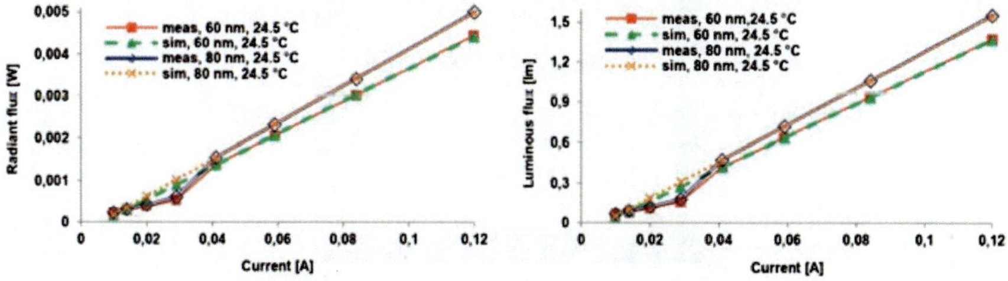

Fig. 8 I-L characteristics: measured versus simulated results (large OLEDs) [12]

$$G_{j\_cell}(I) = \frac{I_{cell}}{\sqrt[m]{\frac{A_{junction}}{A_{cell}} I_{cell}/b}} \quad (3)$$

Electro-thermal and temperature-dependent optical modeling and distributed simulation of large area lighting purpose OLEDs were presented. The extended algorithm of their field solver is able to model large area semiconductor junctions, such as LEP layers of OLEDs properly. The extension can handle any nonlinear function as semiconductor junction equation; the demonstrated electro-thermal OLED model uses power function to describe the junction. The simulator can produce radiance and luminance maps and it can determine the full radiant flux and luminous flux of the light of the simulated devices. The temperature-dependence of luminous flux is significant; in case of their samples it was 5-22% after a 40 °C temperature change, depending on the current [12].

Under thermal boundary conditions, László Poh et al. [13] proposed three natural convection models for vertical convection plates. These models are used in OLED simulators based on electrothermal field solvers. The independent 50 × 50 mm2 active surface OLED is surrounded by still air, and its measured and simulated data will be compared. The structure of the model can be seen in Fig. 9. On the 60 × 60 mm^2 substrate is the effective area of 50 × 50 mm^2. A 5 mm wide metal frame

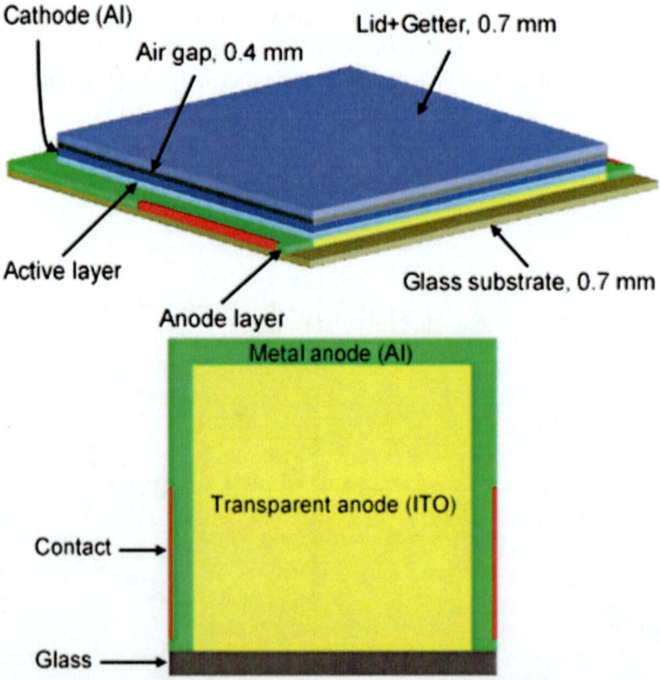

Fig. 9 Simulation model for the 50×50 mm^2 active area OLED. 3D view (not to scale) and the anode layer [13]

surrounds the three sides of the transparent anode, while the other side is covered with insulating material. The sheet resistance of ITO is 20Ω, while the sheet resistance of the frame and cathode is only 0.3Ω. Due to the large difference, the current distribution becomes asymmetric. The current flowing through the surface of the ith cell is

$$I_{cell_i}(U,T) = 100 \times I(U,T) \times \frac{A_{cell\_i\_surface}}{A_{OLED}} \quad (4)$$

Where $A_{OLED} = 2500\ \text{mm}^2$. Their simulation results, measurement results and COMSOL simulation results are shown in Fig. 10. COMSOL results should be similar to their a) simulation results. However, COMSOL results are closer to the measured data. The convection value they calculated was lower than the actual value. Considering the airflow factor, the simulation of their model is better than the COMSOL simulation: the slope of the y contour is larger, and the hottest point is farther from the bottom of the OLED [13].

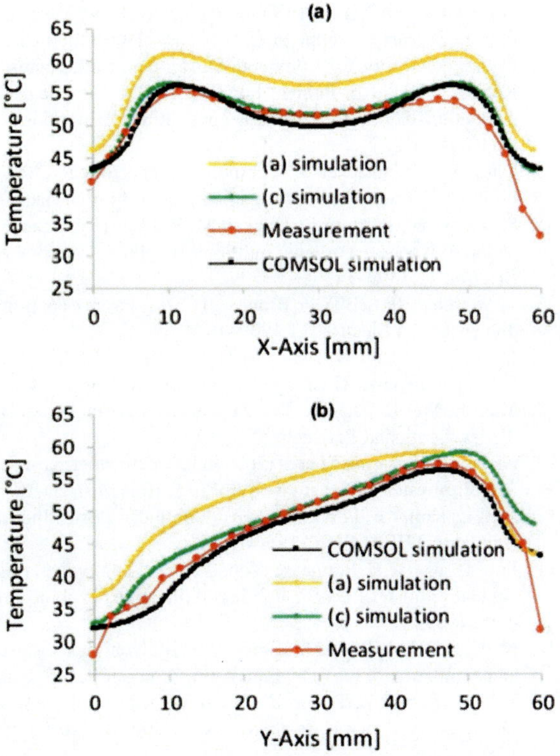

Fig. 10 Midline temperature profiles of the measured [14], COMSOL simulated [14] and the three simulated temperature distributions at the bottom side of the OLED. **a** X-axis, **b** Y-axis

3 Conclusion

This review showcases some of the recent advances in the research of electro-thermal modeling of OLEDs. The simulation model based on finite element, and application of natural convection models for vertical plates in an electro-thermal field solver based OLED simulator, extending of simulation tool SUNRED by using a nonlinear semiconductor junction, approach to distributed electro-thermal field simulation were described and their application demonstrated.

Acknowledgements The authors gratefully acknowledge the Natural Science Foundation of Zhejiang Province Grant No. LY18F050009 for financial support.

References

1. Bender VC, Marchesan TB, Alonso JM (2015) Solid-state lighting a concise review of the state of the art on LED and OLED modeling. IEEE Ind Electron 9:6–16
2. Pohl L, Kohári Z, Poppe A (2017) Fast electro-thermal simulation of large area OLEDs in natural convection environment. Amsterdam, Netherlands 17452719
3. Kohári Z, Kollár E, Pohl L, Poppe A (2013) Nonlinear electro-thermal modeling and field-simulation of OLEDs for lighting applications II: luminosity and failure analysis. Microelectron 44:1011–1018
4. Gärditz C, Winnacker A, Schindler F, Paetzold R (2007) Impact of Joule heating on the brightness homogeneity of organic light emitting devices. Appl Phys Lett 90:103506
5. Slawinski M, Bertram D, Heuken M, Kalisch H, Vescan A (2011) Electrothermal characterization of large-area organic light-emitting diodes employing finite-element simulation. Organ Electron 12:1399–1405
6. Schwamb P, Reusch TCG, Brabec CJ (2013) Passive cooling of large-area organic light-emitting diodes. Organ Electron 14:1939–1945
7. Sturm JC, Wilson W, Iodice M (1998) Thermal effects and scaling in organic light-emitting flat-panel displays. IEEE J Sel Top Quantum Electron 4:75–82
8. Yang L, Wei B, Zhang J (2012) Transient thermal characterization of organic light-emitting diodes. Semicond Sci Technol 27:105011
9. Gielen AWJ, Barink M et al (2009) The electro-thermal-mechanical performance of an OLED: a multi-physics model study. EuroSimE 10647457:26–29
10. Pohl L, Kollár E, Poppe A (2010) Nonlinear electro-thermal OLED modelin SUNRED field simulator. THERMINIC 11654670
11. Pohl L, Kollar E, Poppe A, Kohari Z (2012) Nonlinear electro-thermal modeling and field-simulation of OLEDs for lighting applications I: algorithmic fundamentals. Microelectron J 43:624–632
12. Kollár E, Pohl L, Poppe A, Kohári Z (2014) Nonlinear electro-thermal and light output modeling and simulation of OLEDs. Electri Eng Comput Sci 58(2):43–53
13. Pohl L, Kohári Z, Poppe Z (2018) Vertical natural convection models and their effect on failure analysis in electro-thermal simulation of large-surface OLEDs. Microelectron Reliab 85:198–206
14. Slawinski M, Bertram D, Heuken M, Kalisch H, Vescan A (2011) Electrothermal characterization of large-area organic light-emitting diodes employing finite-element simulation. Org El 12(8):1399–1405